WHY CHIMPANZEES CAN'T LEARN LANGUAGE AND ONLY HUMANS CAN

LEONARD HASTINGS SCHOFF MEMORIAL LECTURES

WHY CHIMPANZEES CAN'T LEARN LANGUAGE AND ONLY HUMANS CAN

HERBERT S. TERRACE

Columbia University Press

New York

Columbia University Press
Publishers Since 1893
New York   Chichester, West Sussex
cup.columbia.edu
Copyright © 2019 Columbia University Press
All rights reserved

A complete cataloging-in-publication record is available
from the Library of Congress.
ISBN 978-0-231-17110-6 (cloth: alk. paper)
ISBN 978-0-231-55001-7 (e-book)
LCCN 2019004823

Columbia University Press books are printed on permanent
and durable acid-free paper.
Printed in the United States of America

For Gillian and Ned

CONTENTS

PREFACE

THIS BOOK is based on the William Schoff lectures I gave at Columbia University.[1] The topic was the evolution of language, words in particular. Darwin argued that language evolved, like all other biological functions, but he offered no insight into its origins. The few ideas that he and other thinkers proposed about bridging the gap between animal communication and language were thought to have such little scientific merit that, after the publication of *The Origin of Species*, discussion of the evolution of language was banned by proclamations of the Société de Linguistique de Paris in 1866 and, in 1872, by the London Philological Society.

That ban, largely honored for almost a century, was broken dramatically by Noam Chomsky's famous review of Skinner's *Verbal Behavior*,[2] a book in which Skinner had claimed that language was a special form of conditioned behavior. Chomsky not only showed that conditioning could not explain language, but he also argued that language was uniquely human and that the gap between animal communication and language was too large to be explained by natural selection.

My interest in the evolution of language began about twenty years later, when I tried to explain the failure of a project in which I attempted to teach sign language to a chimpanzee, Nim Chimpsky. When naming my chimpanzee, I was obviously anticipating a positive outcome. If I could teach Nim to produce sentences in sign language, I would have refuted Chomsky's view that only humans could learn language. Specifically, I would have shown that the ability to create new meanings by combining words—an ability that Chomsky claimed was the defining feature of language—could be found in another species.

It turned out that Chomsky was right but for the wrong reason. Nim couldn't learn sign language, not merely because he couldn't produce sentences but because he couldn't even learn words. Initially, human infants learn such meanings by sharing names conversationally. I concluded that no animal has that ability.

Why were words such an impossible task for a creature as intelligent as a chimpanzee? I addressed that question in my first Schoff lecture by describing various attempts to teach apes to use language, attempts that inspired my own project. I concluded that Nim and similarly trained chimpanzees could learn to use symbols, but only as a means to obtain particular rewards, such as food, drink, tickling, and so on. What they couldn't learn was that things have names.

My second Schoff lecture addressed the origins of a child's first words and their evolutionary significance. Much of that lecture was based on papers from a conference on joint attention and agency that I chaired at Columbia in 2012 (with Janet Metcalfe).[3] Those papers described discoveries by developmental psychologists about nonverbal emotional and cognitive relations that infants form with their caretakers *before* they produce their first words. The absence of that experience in chimpanzees explains why they couldn't learn language.

My second lecture also described recent discoveries by paleoanthropologists about some human ancestors that evolved after our split from chimpanzees, approximately six million years ago. My focus was *Homo erectus* and the circumstances that may have led to the production of the first words. I also explained how the theory of natural selection could account for that remarkable leap, from a limited number of involuntary and immutable signals to the open-ended use of words.

My third Schoff lecture addressed theories of the evolution of language, in particular Chomsky's concept of Universal Grammar. Chomsky is to be admired for transforming linguistics from a static collection of rules for describing individual languages to a quest for a Universal Grammar, one that can generate meaningful sentences in any language. The concept of a Universal Grammar created a revolution in linguistics, but in its present form it has to be refined. Most problematic is its inability to explain the origin of words, which had to be in place before grammar could evolve. By ignoring the origin of words, Chomsky created a serious problem for his theory of language. To create new meanings by combining words into sentences, we have to learn the meanings of individual words.

Universal Grammar also discounts the social function of language, a function I regard as basic for the initial acquisition of the meaning of words. Yet another problem is Chomsky's insistence that grammar emerged abruptly, as the result of a mutation that caused a minor rewiring of the brain about 80,000 years ago. That problem could be avoided by acknowledging that grammar develops in stages, both phylogenetically and ontogenetically. As discussed in chapter 5, that would not alter the basic goal of Universal Grammar, which is to generate a set of well-formed sentences from the input of any language and to discriminate well-formed and ill-formed sentences.

After the failure of ape language experiments, I benefited from discussions about that topic with various colleagues. By happy coincidence, those times coincided with the discovery of many phenomena in related fields that helped to explain the negative results of Project Nim. Two colleagues deserve special mention: Michael Studdert-Kennedy was an outstanding psycholinguist with whom I discussed the evolution of language, during and after Project Nim. Beatrice Beebe is a developmental psychologist and a psychoanalyst who opened my eyes to the significance of an infant's *pre*verbal experience with her mother.

Michael helped me understand the complexities of Chomsky's theories. In 2017, we reviewed Chomsky's book on the evolution of language and explored the strengths and weaknesses of his theories.[4] Beatrice, whom I met in 2005, directs a lab at the Psychiatric Institute of Columbia University, where she uses video analysis to study mother-infant interactions. There she introduced me to her cutting-edge research on intersubjectivity, a significant emotional relation between an infant and her mother that begins at birth and gives way to joint attention, a cognitive relation among an infant, her mother, and external objects. As described in chapter 4, both relations are necessary precursors of an infant's first words.

Michael and Beatrice read and commented on each of the chapters of this book. Other colleagues who read and commented on one or more chapters include Jerome Bruner, Eugene Galanter, Sylvia Goldman, Robert Gordon, Ran Hassin, Tory Higgins, Sarah Hrdy, John McWhorter, Janet Metcalfe, Katherine Nelson, Katheryn Perutz, Robert Remez, Carolyn Ristau, Ann Senghas, and Charles Yang.

While writing this book, I was supported by invaluable advice from the editors at Columbia University Press. That began in 2014 with Patrick Fitzgerald. After he retired, I benefited from comments and suggestions of Brian Smith and Miranda Martin. I am particularly indebted to Adrian Varallyay, an undergraduate student at Columbia University, for his conscientious assistance in preparing figures and copyediting. I would also like to thank the University Seminars at Columbia University for supporting the conference on joint attention and agency and NIMH for supporting some of the research reported in this book (Grant MH111703).

PROLOGUE

FORTY YEARS ago, I published negative results of a project in which I attempted to teach a chimpanzee to use language (Project Nim).[1] That and similar "ape language" projects were in large part motivated by Noam Chomsky's radical and ground-breaking approach to human language. Chomsky not only argued that language was uniquely human but he also questioned Charles Darwin's theory that language evolved from animal communication and B. F. Skinner's theory that language could be reduced to learned behavior.

The ape language experiments challenged Chomsky's view of language by trying to show that chimpanzees (and in one case a gorilla) could create sentences. By the time my project ended, I agreed with Chomsky that language was uniquely human—but I also realized that we were asking the wrong question. Instead of asking whether a chimpanzee could create a sentence, we should have asked whether apes could learn to use words, a simpler task we took for granted.

One of the joys of science is discovering positive implications of negative results. Project Nim is a good example. Little did I realize that Nim's inability to learn language would ultimately lead me to recognize why he failed. This book summarizes some of those reasons. Paleoanthropologists (anthropologists who study immediate ancestors of humans) showed why more recently evolved hominins were more likely to learn words than chimpanzees. Developmental psychologists showed how human infants are uniquely primed to learn words. In light of these discoveries, I examine Chomsky's theory of language and show why, despite his claim otherwise, it doesn't explain the evolution of language.[2] I also show why his theory is limited to language as it is currently used.

■ ■ ■

My initial connection with Chomsky was indirect. While working with Skinner as a graduate student at Harvard, I read Chomsky's scathing review of Skinner's theory of language.[3] At the time, I was doing research for my doctoral dissertation on teaching pigeons to discriminate stimuli without making errors—a far cry from language. Although I had an abiding interest in language, I continued to focus on my dissertation research. Because Skinner was one of the most influential and persuasive psychologists of the twentieth century, I assumed that he would respond to Chomsky's challenge, as he had to many earlier criticisms of his theories. To my chagrin, he did not.

After obtaining my PhD, I joined the faculty of Columbia University, where I continued to study discrimination learning in pigeons. However, I was often distracted by reports claiming that chimpanzees had learned language.[4] If those reports were valid, they would not only provide evidence of continuity between animal communication and language but would also show that some features of language could be conditioned.

Early attempts to train apes to speak failed,[5] ostensibly because apes lacked the articulatory apparatus to produce the sounds humans make when they speak. More recent ape language projects introduced two ingenious methods to overcome that limitation. One trained chimpanzees to learn American Sign Language (ASL), a gestural language used by hundreds of thousands of hard-of-hearing people in the United States.[6] Instead of spoken words, sign language uses gestures.

Like spoken words, the vast majority of signs are arbitrary. This means that a "listener" can't discern the meaning of the signs from their physical characteristics. Just as there is nothing about the sound of the spoken word *woman* that implies a human female, there's nothing about the gesture of a person scratching his cheek with a thumb to convey the same meaning in ASL. That sign was created during the nineteenth century, when women wore hats that were secured by strings tied under the chin. Similarly, the sign for *man*—the tips of one's fingers brushing the side of the head—was inspired by the brims of hats that were a common feature of a man's wardrobe at that time. Styles have since changed but signs haven't, hence their opaqueness and arbitrary nature. Furthermore, just as people in other countries use different words to refer to a particular object, deaf people do the same with different signs.

The second method for overcoming the vocal limitations of chimpanzees was to use visual symbols that refer to particular objects or events. Here again, the meaning of each symbol is arbitrary. For example, on one project, a blue triangle meant *apple*, and a black star, *insert*.[7] By combining sequences of different symbols, chimpanzees were trained to make particular requests—for example, *play music*.[8]

The more I read about the ape language projects, the more it seemed as if they might actually succeed. Particularly intriguing was a report that Washoe, the first chimpanzee to learn sign language, had combined two gestures to produce a specific meaning—for example, the signs *water* and *bird* to describe a swan. Roger Brown, who at the time was the leading authority on language acquisition by children, reacted by saying, "It was rather as if the seismometer left on the moon had started to tap out 'S-O-S.'"[9] Brown thought that Washoe's combinations were evidence that a chimpanzee could create a sentence.

If so, I envisioned the possibility of having a conversation with a chimpanzee, such as asking a linguistically trained chimpanzee to tell me about natural communication between other chimpanzees. That would not be terribly different from the plot of the Planet of the Apes films.[10] Curiously, those films began to appear with some regularity a few years after the ape language projects were begun.

I was troubled, however, by the anecdotal quality of reports of language use by chimpanzees. They reminded me of the cliché that "the plural of anecdote is not data." To confirm the grammatical ability of a chimpanzee, I thought it was necessary to obtain what psycholinguists refer to as a "corpus," a collection of *all* of the utterances a child makes and the circumstances under which they occurred.[11] Seeing no evidence of such an undertaking, I decided to start my own project. In 1972, I initiated one centered around a two-week-old male chimpanzee who was born at a primate center in Oklahoma. After he was flown to New York, I named him Nim Chimpsky.

Within thirty months, members of Project Nim assembled a corpus of more than 20,000 combinations containing two or more of his signs. Initially, I thought the corpus provided enough evidence that a chimpanzee could create a sentence. But as I was about to publish my results, I made a shocking discovery. While watching a videotape of Nim signing with one of his teachers, I noticed that she inadvertently cued most of his signs.

I had used that and similar tapes to acquaint new teachers with the methods we used to teach Nim to sign. Unlike other viewings, in which I only watched Nim, on this occasion I watched both participants. What I saw was heartbreaking. In anticipation of Nim's signing, his teacher unwittingly signed the sign(s) she expected him to sign. About a quarter of a second later, he made one or more of those signs and was given a reward.

Instead of naming the objects with which he played, Nim signed merely to obtain a reward. Videotapes and films of other apes who had been trained to use sign language revealed the same pattern of prompting by a trainer and reward for signing.[12] Under similar circumstances, a child would *name* the objects she played with in order to obtain praise from her parents. No physical reward is needed.

Analyses of the performance of chimpanzees trained to use visual symbols also led me to conclude that their responses were meaningless. Although prompting could not explain their performance, the combinations of symbols they produced were acquired during laborious sessions of rote training. Sequences produced by chimpanzees did not differ from the meaningless passwords that people use to obtain cash from an ATM.

WORDS BEFORE GRAMMAR

Although I agreed with Chomsky that chimpanzees couldn't learn language, there were other issues about which we didn't see eye to eye. One was his emphasis on grammar at the expense of the word. Another was his claim that the primary function of language was to help us think, rather than to share knowledge.

In hindsight, it's easy to understand Chomsky's position about language. He began his brilliant career by calling attention to language's richest feature: a grammar that allowed a speaker to create an infinite number of meanings from a finite vocabulary. No other system of communication has that feature. Add to that his theory that all languages are governed by a Universal Grammar —one that allows the speaker of one language to readily learn any other—and it's easy to see why Chomsky's theories dominated linguistics for more than sixty years.

Everyone agrees about the importance of grammar in defining language, yet words don't deserve the back seat they've been allotted by Chomsky.[13] Language and animal communication clearly differ, but they differ in two respects. Animals lack not only grammar but also words. And words had to evolve before grammar.

To explain how a child learns language, Chomsky proposed a language acquisition device (LAD), an innate neural mechanism that enables the child to learn the rules of her grammar without explicit feedback from her elders. The LAD attempted to explain how children master grammatical irregularities—for example, saying *went* instead of *goed*. The LAD cannot, however, explain the acquisition of words.

THE PRIMARY FUNCTION OF LANGUAGE

Chomsky argues that the primary function of language is thinking. That position reflects his criticism of Skinner's view that language is shaped by the environment and Skinner's insistence that the core features of language were determined by biological rather than psychological factors.

I disagree. The only way a child can learn language is through *conversation*, a communicative act. Claiming that the main function of language is to help us think takes for granted years of education in which a child is taught how to think. It's like theorizing about building a skyscraper without knowing how to build the simplest type of shelter or theorizing about making airplanes without knowing how to fly a kite.

Linguists and psychologists have yet to agree on what steps are needed to learn language, but after Project Nim ended,

I encountered two important clues. One was suggested by paleoanthropologists; the other by psychologists who study emotional and cognitive precursors of language in infants.

In the 1970s, paleoanthropologists began to discover fossils of human ancestors who evolved after hominins split from chimpanzees about seven million years ago. Their discoveries not only provided a firm foundation of recent human evolution but also showed that chimpanzees were more remote in our history than they appeared to be when the ape language experiments began. Paleoanthropologists also discovered how our ancestors adapted to changes in climate and suggested a species that may have produced the first words.

Shortly after Project Nim ended, developmental psychologists began to discover emotional and cognitive interactions between human infants and their mothers during an infant's first year. Those interactions are necessary antecedents of an infant's first words. Their absence in apes seems to be the best explanation of why they can't language.

Years after Project Nim ended, it became abundantly clear that apes were unable to learn language because they couldn't learn that things have names. Paleoanthropologists suggested that human ancestors, who were more cooperative than chimpanzees, may have learned to use names to motivate their peers to scavenge for meat. Developmental psychologists showed that infants have to develop strong nonverbal emotional and cognitive bonds with their caretakers before they can learn to produce their first words.

Taken together, these secular discoveries support the biblical assertion: "In the beginning was the Word."

WHY CHIMPANZEES CAN'T LEARN LANGUAGE AND ONLY HUMANS CAN

Chapter One

NUMBERLESS GRADATIONS

ABOUT 3.8 billion years ago, the earth was just another lifeless speck in an infinite universe. Since that time, more than eight billion species of plants and animals have lived on our planet. In 1859, Charles Darwin proposed an ingeniously simple theory to account for this diversity.[1] He suggested that all existing species evolved by descent from a common ancestor by a process called *natural selection.*

Without any constraints, the size of the population of each species would increase exponentially. This did not happen because of steady competition for a finite set of resources in unpredictable environments. The prevailing rule was "Nature, red in tooth and claw."[2] Because of predation or disease, some members of a species died before they could reproduce. The offspring of surviving members of any given species engaged in a new cycle of competition for survival. During that process, variations of genetic structure occasionally led to the formation of a new species. It is generally agreed that successful breeding is the criteria for distinguishing old and new species. Members of a new species cannot produce viable offspring by mating with members of an old species.

Thomas Huxley, one of Darwin's strongest supporters, remarked, "How extremely stupid not to have thought of that!"[3] Many other scientists agreed. During the following century, thousands of publications confirmed predictions of the theory of evolution. In 1973, Theodosius Dobzhansky, an eminent twentieth-century geneticist, concluded that "nothing in biology makes sense except in the light of evolution."[4]

Language is the major exception. The theory of evolution has yet to explain it. Darwin tried, but his strongest statement about its origin was vague: He argued that language evolved from simpler forms of animal communication during the course of "*numberless gradations*" (italics added).[5] We have yet to discover what those gradations are.

The challenge is to fill the seemingly unbridgeable gap between animal communication and language. As Noam Chomsky famously asked, how could language, a voluntary form of communication, have been selected from the involuntary grunts and screams of chimpanzees, our closest living ancestors?

Some of Darwin's contemporaries also thought that the gap between animal communication and language was too large. For example, Max Müller, a professor of linguistics at Oxford, asserted that:

Language is the Rubicon which divides man from beast, and no animal will ever cross it . . . the science of language will yet enable us to withstand the extreme theories of the Darwinians, and to draw a hard and fast line between man and brute.[6]

Experiments on "ape language," which I describe in chapter 2, support Müller's conclusion that animals cannot acquire language. But he is wrong about evolutionary theory. There is simply no scientific alternative. What remains elusive are the numberless gradations that led from animal communication to language.

Some members of the European intelligentsia, Darwin included, naïvely argued that our ancestors' first words were imitations of natural sounds: for example, the sounds of a dog barking, a snake hissing, a cat purring, a duck quacking, workers exerting themselves, and so on.[7] In this view, language was just another example of animal communication, albeit more complex. The obvious and fatal problem with such "onomatopoetic" theories is that, even if true, they could account for only a miniscule portion of vocabulary. In response to the facile quality of the theories being offered, Société de Linguistique de Paris (in 1866) and the London Philological Society (in 1872) banned all further discussions of the origin of language shortly after the publication of the *Origin of Species*.

That did not deter a strong desire to alleviate what many thinkers regarded as the Achilles' heel of the theory of evolution. Following Darwin, linguists, anthropologists, psychologists, neuroscientists, biologists, computer scientists, philosophers, and others have written about the evolution of language. This book continues that effort but with a novel approach. Instead of focusing on grammar, our most powerful linguistic ability, I ask a simpler but equally important question: Why are humans the only species that uses words, that knows that things have names, and that can use names conversationally?

Why opt for words instead of grammar? Quite simply, I changed my mind about their relative importance after the failure of Project Nim, the goal of which was to teach a chimpanzee to create a sentence. The failure of that project led me to realize that chimpanzees were not only unable to produce sentences, but they were also unable to produce words. I recognized that asking how language evolved was actually asking two questions: one about the origin of *words*, the other about the origin of *grammar*.

Understanding the origin of words is a much more difficult problem than it might first appear. Like walking, talking is so universal that it hardly seems worthy of our attention. Indeed, walking and talking are similar in that, with no obvious instruction, humans seem to develop both abilities on their own.

But they don't. To see why not, imagine the proverbial story of a newborn infant who is raised alone on an island on which all of her basic needs are met. In theory, a robot could be programmed to satisfy those needs (for example, to provide food, drink, warmth, and so on). Would this infant learn to walk? Of course! But would she learn to talk? Other than the reflexive cries one would expect from an infant in distress, I can't think of any reason that she would. And even if that infant exclaimed her needs, and had them satisfied, there's no reason to think that she would be able to conduct the simplest type of conversation.

The moral of this thought experiment is that learning to walk is a product of maturation. Before an infant can learn to talk, however, she must experience certain affective and cognitive interactions with others, usually her mother. Those interactions are nonverbal and uniquely human. Affectively, an infant learns to share her emotions. Cognitively, she learns to share her perceptions of interesting events. Both types of sharing are conversational. When viewed from this perspective, the evolution of language depends on the evolution of the ability to converse nonverbally and verbally.

There's no evidence that a nonhuman animal can engage in conversation. The signals that animals use to communicate encompass a variety of physical dimensions. Male primates often stare at other members of their group to assert their position in a dominance hierarchy.[8] Some species release pheromones to attract mates.[9] Certain varieties of fish generate electrical signals during courting.[10]

Birds sing to attract mates and to assert their territory. Superficially, it may seem that songbirds' duets are conversational, but analyses of their duets have shown that the songs are innate and that the topic of their exchanges does not vary.[11] A songbird's message doesn't transmit any new information. As another example, foraging bees communicate the location and quality of nearby food to their hive mates by engaging in elaborate "round" and "waggle" dances. Those dances, however, are not learned and are immutable. A bee cannot substitute a new dance to communicate the source and the quality of nearby nectar.[12] The same is true of the distinctive alarm calls that vervet monkeys produce when they perceive particular predators.[13]

However fascinating, such innate signals are few in number and rarely exceed thirty different types in a particular species.[14] By contrast, human mothers and infants engage in dialogues that reflect how they each perceive the other's emotional and mental state.

Foolishly, I never considered the role of conversation when I began Project Nim. It was generally assumed that all communication was conversational, including the communication described in the initial reports of ape language experiments. That assumption, which was clearly wrong, reflected my then anticognitive bias. Having been trained as a behaviorist, I was taught *not* to ask questions about the emotions and mental states of a speaker and a listener.

Psychologists are still struggling to answer these questions; but, as we shall see in chapter 4, progress has been made by supplementing behavioral methods rather than by replacing them. To understand those approaches, we must first understand the origins and limitations of behaviorism, the dominant theory of psychology during the first half of the twentieth century. For that, we must digress to the beginning of comparative psychology.

COMPARATIVE PSYCHOLOGY

In 1865, Herbert Spencer, a disciple of Darwin, broadened the theory of evolution by arguing that natural selection applies to *behavior* with the same logical force that it does to anatomical structures.[15] Just as a creature's sensory organs, brain size, or skeleton can be said to be adaptive, so can particular types of behavior. For example, a species can be distinguished by the degree to which its behavior is instinctive (unlearned) or the degree to which a particular behavior is susceptible to different types of classical or operant conditioning.

Spencer's insight gave rise to *comparative* psychology, an area of psychology that examines similarities and differences in the behavior of different species. For obvious reasons, comparative psychologists were also called *behaviorists*. Of particular interest were similarities between human and primate intelligence that pertain to the evolution of language.

Research on insight in apes provides an instructive example. In a famous experiment by Wolfgang Köhler, Sultan, a captive chimpanzee, was faced with the problem of obtaining a banana that hung from the ceiling of his cage. To obtain the banana, Sultan stacked some wooden boxes that he played with, and then stood on top of the stack to reach the banana. Because he was not explicitly trained to do so, Köhler described Sultan's "insight" as a remarkable intellectual achievement, one that deviated sharply from trial-and-error learning.[16]

Most comparative psychologists reject insight and other mentalistic concepts because evidence of those concepts cannot be observed directly. Instead, they base their theories exclusively on objectively defined stimuli and responses, a credo of behaviorism.

Many years after Sultan became a textbook example of an intelligent ape, a replication of Köhler's experiment showed that Sultan's performance actually required trial-and-error learning. Two groups of chimpanzees were trained under conditions identical to those used with Sultan.[17] For one group, no boxes were available until the test. For the second, the boxes were a permanent feature of the subjects' home cage, as in the original experiment. "Insightful" solutions were obtained only by the latter group, the one that had had ample opportunity to play with the boxes. This experiment showed that Sultan's "aha" moment could be explained by his experience and familiarity with stackable boxes.

CLASSICAL CONDITIONING

At the beginning of the twentieth century, behaviorists used two methods to study learning. One focused on involuntary behavior, the other on voluntary behavior. *Involuntary behavior* is behavior that can be elicited regardless of its consequences; an example is salivation in response to ingestion of food. *Voluntary behavior* is behavior, the frequency of which can be modified by its consequences; an example is a rat pressing a bar for food. In a classic experiment, Ivan Pavlov showed how a new involuntary reflex could be trained by using a neutral stimulus to signal an established response.[18] First, he signaled the presentation of food by ringing a bell. Initially, the dog ignored the sound, but after a few pairings of the food and the bell, the dog began to salivate each time the bell was rung.

In this brilliant feat of biological engineering, Pavlov extended the domain of a natural reflex (food → salivation) by showing how to create a new reflex (bell → salivation). But Pavlov's achievement was limited because it made no provision for training *new* behavior. Pavlov's dogs simply learned to make the involuntary response of salivating (an old trick) under new circumstances. Another limitation was the fact that conditioned salivation was involuntary. A reward was given whether or not conditioned salivation occurred.

INSTRUMENTAL CONDITIONING

About forty years later, B. F. Skinner showed how to condition voluntary behavior by using a remarkably simple procedure. He delivered a reward after the desired response occurred, or at least some approximation of the desired response.[19] He referred to this type of conditioning as *instrumental* because a reward was contingent on the occurrence of a particular response. Skinner argued that instrumental conditioning was the basis of a huge variety of everyday behavior.[20] Students seek high grades by writing good essays; people control their body temperature by adding or shedding clothing; they learn to lie to avoid shame and to drive slowly to avoid speeding tickets.

Skinner explained such changes in behavior by appealing to a contingency between a response and its consequence. As shown in table 1.1, a response can produce, avoid, or escape from a particular reward. Those rewards could be primary or secondary, positive or negative, or some combination thereof.

Primary rewards are based on innate needs, such as food, warmth, or the elimination of a noxious stimulus. Secondary rewards are based on the association between a new stimulus and a primary reward: for example, a tone or money that is

TABLE 1.1 Examples of contingencies of reinforcement

Response	Consequence	Type of Reward	Contingency
Picking berries	Food	Positive primary	Reward training
Doing chores	Praise	Positive secondary	Reward training
Writing a good essay	Good grade	Positive secondary	Reward training
Hugging a parent	Affection	Positive primary	Reward training
Putting on a fur coat in cold weather	Warmth	Positive primary	Reward training
Telling a lie to avoid blame	Social acceptance	Positive secondary	Avoidance learning
Postponing a dentist appointment	Delay of pain	Positive secondary	Avoidance learning
Touching a hot stove	Pain	Negative primary	Punishment
Getting caught stealing a cookie	Time out	Negative secondary	Punishment
Getting a speeding ticket	Loss of money	Negative secondary	Punishment

paired with food or other primary rewards, a smile that is followed by a hug, a green light that signals when it's safe to cross the street, and so on.

Having demonstrated the power and the versatility of a contingency between a response and its consequence, Skinner noted that contingencies are seldom learned independently of their context. If, for example, you want to obtain candy from a vending machine, the vending machine has to be plugged in (lit and powered). Otherwise, inserting a coin won't produce a chocolate bar. Similarly, a gate has to be open before you can drive through it. In these examples, neither the powered vending machine nor the open gate elicit behavior. They simply set the *occasion* for a particular contingency to be in effect.

The addition of an occasioning stimulus—or as it is technically called, a *discriminative* stimulus—greatly broadens the scope of instrumental conditioning. For example, a friend's smile might indicate that it's okay to approach her; a scowl, that it's not. Skinner also noted the role of discriminative stimuli in verbal learning. He argued that factual knowledge can be described by the relation between a verbal (discriminative) stimulus and a particular response that is rewarded by a member of the social community. For example, in the presence of the discriminative stimulus 2×3, the response, 6, earns a point on an exam. The same would be true for the response *water* in the presence of the discriminative stimulus H_2O, or the response *red* in the presence of the discriminative stimulus *rouge*.[21]

Skinner's ingenious ability to analyze and organize voluntary behavior into particular combinations of discriminative stimuli and instrumental responses enabled him to develop a research program that appeared to have no limits. Pigeons, for example, were trained to "guide" missiles aimed at enemy targets[22] and to discriminate various kinds of concepts, exemplars of which were projected onto a translucent surface that could detect their pecks.[23] Monkeys were trained to solve logical "oddity" problems by selecting the odd stimulus from a trio of geometric forms.[24] "Teaching machines" trained people to learn a variety of technical facts or the vocabulary of a foreign language.[25] Clinical psychologists used the principles of instrumental conditioning to modify the behavior of their patients,[26] as did teachers who modified the behavior of unruly high school students.[27]

Skinner also recognized that serially organized action is the norm in everyday behavior, as compared to individual responses. This idea motivated a model that integrated combinations of discriminative stimuli and instrumental responses into *chains* of behavior. Skinner argued that such chains are fundamental for the mastery of skills at all levels of complexity, whether those skills are as simple as knowing how to get from one place to another or as complex as using a language.

Consider, for example, how you might find your way in a new neighborhood. When you start, you're likely to make many errors, such as turning the wrong way at a particular intersection. If each intersection is thought of as a discriminative stimulus (say, one with a statue, one with a bank, one with a church, and so on), the task of finding your way boils down to learning which way to turn when encountering different discriminative stimuli.

In this model, each discriminative stimulus actually serves two functions. It provides a secondary reward for the behavior that led to its appearance (for example, walking east on Main Street). It also functions as a discriminative stimulus for making the next turn (walking down Jones Street toward the church), and so on.

In 1957, Skinner attempted to show how a chaining model could explain language. In *Verbal Behavior*,[28] he argued that language was simply a collection of verbal habits that children learned by trial and error and/or by imitating their caretakers' utterances. He also argued that sentences were constructed one word at a time. For example, Skinner would explain a sentence such as *Bill loves Mary* as a chain of conditioned responses. First, the speaker selects *Bill* from a list of x proper nouns, then *loves* from a list of y verbs, and finally *Mary* from a list of z proper nouns.

CHOMSKY'S CRITIQUE OF *VERBAL BEHAVIOR*

Behaviorism dominated psychology for the first half of the twentieth century. Research in areas as diverse as human memory, psychophysics, perception, and animal learning assumed that behavior was the *only* dependent variable in psychology, whether the behavior was a response on a paired-associate test, the detection of some stimulus at threshold, the perception of a figure from ground, or the planning of a route from point A to point B.

It's no exaggeration to say that Chomsky's review[29] of Skinner's *Verbal Behavior* was one of the most influential and devastating critiques of behaviorism, and one that helped to create the field of cognitive psychology. Chomsky's review showed why language was more than a chain of conditioned responses. More generally, he showed why behaviorists couldn't account for the ability to create an essentially infinite number of new and meaningful sentences by combining a finite set of words.[30] For example, a child who learns to say the declarative sentence *There's Bill* can readily convert it into an interrogative sentence, *Is that Bill?*, without any explicit instruction.[31] Or the child might transform the declarative sentence *Bill loves Mary* into such variations as *Mary is loved by Bill, Bill doesn't love Mary, Will Bill love Mary?, Bill hadn't loved Mary*, and so on.[32]

The transformation of declarative sentences into other grammatical sequences illustrates the distinction Chomsky drew between *competence* and *performance*. He argued that performance always underestimates competence. For Chomsky, performance is only the tip of the iceberg of grammatical knowledge, although performance is the only thing a behaviorist can measure. Chomsky and other cognitive psychologists were concerned with the knowledge that makes particular types of performance possible.

Embedded sentences provide another example in which performance underestimates knowledge. Consider the following sentence, in which a father tells his son that *Ted Williams, who spent a lot of time fishing with custom-made fishing rods and who was also a famous pilot during World War II, had the highest seasonal batting average in baseball*. Despite the gap of twenty-three words between the noun *Williams* and its predicate *had*, the son would have no trouble knowing the relation between those words, even if one or more additional phrases were embedded. Behaviorists cannot explain relations between nonadjacent words because they assume that the only influence on each word was the previous word. That problem becomes progressively worse as the number of intervening words increases.

Chomsky also showed that chaining theory could not even account for the meaning of adjacent words, as in ambiguous sentences and phrases such as *they are visiting firemen* and *the shooting of hunters*. A nonsensical sentence that Chomsky invented illustrates the reverse problem: *green ideas sleep furiously.* That sentence is grammatically correct but meaningless.

The rapid and effortless acquisition of vocabulary and a child's untutored knowledge of grammar led Chomsky to conclude that children were born with a language acquisition device (LAD), an innate neural mechanism that implemented the rules of a Universal Grammar (UG) that could account for the generation of sentences in *any* of the more than six thousand languages that people speak. The LAD enables children to learn the richness and the complexity of their native grammars without explicit instruction.[33]

Given the austerity of the behaviorist zeitgeist that prevailed when he published his groundbreaking analysis of grammar,[34] Chomsky's anti-Darwinian stance in the 1960s is not surprising. In that context, his insights about the emptiness of stimulus-response explanations of language were a monumental achievement. They did not, however, address the evolutionary origins of language. By focusing on grammar, Chomsky inadvertently magnified that problem. Language and animal communication differ in two important respects. Human discourse is not only grammatical, it is also *conversational.* Whereas children have to learn words before they learn grammar, Chomsky's only concern was grammar.

Chomsky and other linguists asked, quite reasonably, how UG could have evolved by natural selection during the mere six to seven million years that followed the separation of chimpanzees and the human species—an eye blink in evolutionary time. Furthermore, if grammar did evolve in stages, Chomsky asked, what kind of grammars would those stages entail? Of what use, he asked, would half a grammar be, and which portion of language would be included in that half?

However, a moment's thought would show that *any* fraction of a full grammar would be adaptive. For example, the likelihood of survival of an ancestor who could only use the present tense might not be as great as that of someone who could also use the past or the future tenses, but that ability would still add to that its survival value.

I applaud Chomsky's quest for a Universal Grammar as a brilliant strategy for understanding language as it is *currently* used. How else may we explain the facts that children go through the same phases while learning any language, and that the meanings in one language can be shared effortlessly with the meanings of the thousands of other languages that humans use? The concept of UG should not, however, constrain the nature of its origins.

In justifying the significance of UG, Chomsky went one step too far. Because of the vast differences between animal communication and language, he concluded that the evolution of language could *not* be incremental.[35] Language seemed too complex to be derived from animal communication, but that's because Chomsky didn't recognize the evolution of *words* as a separate stage in the evolution of language. As Darwin might say, he didn't recognize that the evolution of words was one of the many numberless gradations needed to bridge the gap between animal communication and language.

Chomsky's concern about the complexities of language are reminiscent of arguments against the theory of evolution that were based on the complexity of the structures it sought to explain. Consider, for example, the question that Darwin posed about the evolution of the human eye:

> To suppose that the eye, with all its inimitable contrivances for adjusting the focus to different distances, for admitting different amounts of light, and for the correction of spherical and chromatic aberration, could have been formed by natural selection, seems, I freely confess, absurd in the highest possible degree.[36]

But upon reflection, Darwin, the quintessential gradualist, formulated a strategy for unraveling that complexity:

> Yet reason tells me, that if numerous gradations from a perfect and complex eye to one very imperfect and simple, each grade being useful to its possessor, can be shown to exist; if further, the eye does vary ever so slightly, and the variations be inherited, which is certainly the case; and if any variation or modification in the organ be ever useful to an animal under changing conditions of life, then the difficulty of believing that a perfect and complex eye could be formed by natural selection though insuperable by our imagination, can hardly be considered real.[37]

Darwin's faith in natural selection was confirmed by biologists who studied how the eyes in different species react to light and how that mechanism gradually became more complex. In each instance, light energy is first transduced into chemical energy and then into electrical energy. Electrical energy is then used by the central nervous system to determine the nature of the stimulus. Figure 1.1 shows some examples of the structure of the eye at different stages of its evolution.[38]

In the case of mental skills, Darwin suggested a similar approach:

> We must also admit that there is a much wider interval in mental power between one of the lowest fishes, as a lamprey or lancelet, and one of the higher apes, than between an ape and man; yet this interval is filled up by *numberless gradations*.[39]

Could that strategy explain the evolution of language? Much more information would be needed to decide. Unlike the transduction of light, which is involuntary, language is voluntary. We also lack intermediate fossils and any knowledge of where in the brain we should look for evidence of earlier stages of language. More problematic is the fact that, unlike vision, which can be fully characterized in an individual, language requires the coordination of the cognition and the behavior of two individuals, a speaker and a listener. These problems notwithstanding, we should keep in mind that complexity should not discourage evolutionary inquiry. As Darwin remarked, the challenge is to define and identify the relevant intermediate steps.

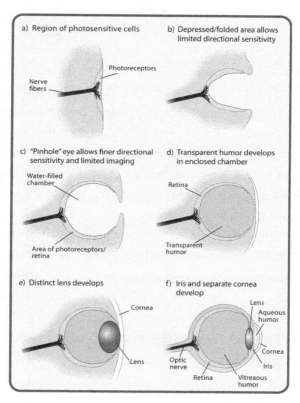

FIGURE 1.1 The evolution of the eye. Complex eyes evolved over many millions of years from a simple photosensitive cell. Eyes vary in their acuity, the range of wavelengths they can detect, their sensitivity in low light levels, their ability to detect motion or resolve objects, and their ability to discriminate color.

Wikimedia Commons.

■ ■ ■

At this point, we should take stock of the debate between Skinner and Chomsky about the nature of language and its evolution. As did Darwin, Skinner adopted a mechanistic view of language. He sought to explain it by the principles of operant conditioning: in particular, his theory that sentences were sequences of conditioned responses. Chomsky showed why that approach wouldn't work, but he also argued that the theory of evolution couldn't explain language, at least by the principles of natural selection.

Chomsky's view did not sit well with behaviorists who believed that language had to have evolved by natural selection. Even though they could not answer Chomsky's theoretical objections, they were strongly influenced by Skinner's behavioral sophistication and his unequaled ability to apply the principles of operant conditioning to all kinds of behavior, both animal and human.

APE LANGUAGE PROJECTS

It was therefore not surprising that, in the 1960s, various behaviorists independently began projects the goal of which was to show that apes—chimpanzees in particular—could learn the rules of a simple grammar.[40] Those projects provided the strongest possible tests of any linguistic overlap between humans and another species.

Although previous efforts to teach an ape to speak had failed miserably,[41] some attributed those failures to the chimpanzee's inability to articulate human sounds.[42] Indeed, it was hypothesized that earlier failures could be overcome by shifting from auditory to visual languages.[43] One approach was to train chimpanzees to use the signs of American Sign Language (ASL), a natural gestural language. The hope was that they would spontaneously combine some of the signs into simple sentences. Another approach trained chimpanzees to use visual symbols and then to combine those symbols into sentence-like sequences.

Initially, some of these projects suggested that nonhuman animals could learn simple grammars. If true, such claims would question traditional criteria for distinguishing between humans and animals. However, as I elaborate in chapter 2, both approaches failed. There was no evidence that chimpanzees knew the meanings of the signs they learned. Although they accumulated sizeable "vocabularies," video analyses of their signing and use of symbols showed that they were motivated by rewards and not by sharing knowledge. In experiments using signing, trainers unwittingly prompted most of the chimpanzees' signs, usually a fraction of a second before the signs were uttered (made). The sequences of visual symbols that chimpanzees learned to produce had no more meaning than sequences learned by rote.

None of the sequences that chimpanzees learned to produce required a linguistic explanation. In each instance, chimpanzees learned to use signs or symbols as imperative commands: that is, as a means of obtaining rewards they could not otherwise obtain. Although human language has imperatives, their relative frequency is miniscule. The overwhelming majority of words are declarative, and their sole function is to share information.

We are, therefore, left with the same issues we faced before the ape language projects began. Language appears to be uniquely human, but we still lack a theory about how it could have evolved by natural selection. Thanks to two new lines of research that emerged after the ape language projects ended, that state of affairs has begun to change.

Paleoanthropologists provided evidence of human ancestors who evolved after we split from chimpanzees. Their discoveries not only showed how evolutionarily remote we are from chimpanzees, but, as I discuss in chapter 3, they also suggested a species that may have been the first to use words, *and* how natural selection could explain the occurrence of words. Developmental psychologists discovered that the emotional and cognitive bonds that an infant forms with her mother are crucial precursors of the infant's first words. Those precursors, which are uniquely human, are discussed in chapter 4.

FROM CHIMPANZEES TO HOMININS

Paleoanthropologists have discovered more than two dozen human ancestors who evolved after hominins split from chimpanzees about six million years ago. Climate change in East Africa was a major contributing factor. The temperature dropped dramatically and became too low to support trees and the many forms of vegetation on which apes thrived. In many areas, trees were replaced by open grasslands.

The skills needed to cope in that environment required a new type of locomotion. In forests and jungles, apes climbed trees quadrupedally (that is, using all four limbs), and swung from one branch to another when moving between food sites. Without the shelter of trees, it was necessary to travel bipedally across open plains. That allowed them to have a better view of their surroundings, in particular, of prey. It also freed them to use their upper limbs to carry food and supplies. Yet another advantage of bipedalism was a reduction of body temperature. When moving quadrupedally, our ancestors bore the full brunt of the sun's rays. Bipedal posture minimized the area that was directly exposed to the sun.

Bipedalism was the first of two major anatomical changes that distinguished our ancestors from chimpanzees. The second was an increase in brain size. That hardly changed during the first three to four million years after our ancestors became bipedal. However, starting with *Homo habilis*, the first ancestor to assume a fully upright posture, brain size began to increase at an accelerating rate.

About three million years ago, our ancestors began to use their hands to manufacture stone tools. With larger brains, our ancestors also learned to cooperate to a much greater degree than chimpanzees would. They began to use tools and to scavenge for meat, and, for the first time, they began to emigrate from Africa.

With one exception, bipedalism had no direct influence on the evolution of language. The exception was a reduction in the size of the pelvis. As our ancestors became bipedal, the size and functions of the pelvis changed. It shrank, and with it the birth canal. For millions of years, a smaller pelvis had no consequence. But as brain volume increased, the size of the birth canal limited the size of an infant's brain. Once brain volume exceeded 1,000 cc, an infant's head could no longer pass through the birth canal without fatal damage to either mother or infant or both. Because of that constraint, the initial volume of a human infant's brain is only 20 to 25 percent of its adult size. By contrast, the volume of an infant chimpanzee's brain is 40 to 50 percent of its adult size.[44]

Because of their poorly developed skeletal and muscular systems, human infants required much more maternal care than did other newborn primates. Whereas most nonhuman infants can wander away from their mothers by the time they are one month old, human infants are not even able to crawl until they are six months old. That's why human infants have to be cradled. Cradling had an unexpected advantage. In compensation, as it were, for the infant's lack of mobility, infant and mother can observe and anticipate each other's behavior during cradling. The mutual gaze between infant and mother that cradling affords is uniquely human.

In sum, there are at least three reasons to consider why recent hominin ancestors provide a more realistic baseline than chimpanzees for clues about the evolution of language: bipedalism, a large brain, and a small birth canal. Although none of these differences were selected to enhance linguistic ability, the need to cradle a human infant for six months led to profound changes in maternal care. As I elaborate in chapter 4, that care played a crucial role in the development of language.

DEVELOPMENT OF NONVERBAL LANGUAGE DURING
AN INFANT'S FIRST YEAR

Shortly after birth, a human infant and her mother develop a bond that is based on reciprocal expressions of affect. Initially, that bond was similar to the attachment that takes place between primate infants and their mothers,[45] but it soon became much more elaborate. Human infants and their mothers engage in *rhythmic* exchanges of vocalization, looking, touching, laughing, smiling, moving their heads, and other emotional responses.[46] Beginning at about six months, the infant and her mother begin to share attention to external objects. The first relation is referred to as *intersubjectivity*;[47] the second as *joint attention*.[48] Both relations are nonverbal and uniquely human.

Intersubjectivity: To document intersubjectivity, developmental psychologists devised new methods for measuring emotional communication. For example, a three-month-old infant and her mother are seated, facing each other, and are videotaped by separate cameras. The videotapes are then time-synched. Sharing of affect between mother and infant is assessed by independent observers who analyze the tapes during normal and slow-motion playback.

Such analyses have shown that infant and mother coordinate their affect and activities. That allows them to predict each other's behavior. For example, a micro-analysis of vocalization measured the onset and offset of a mother and her infant's vocalizations and pauses.[49] On average, mother and infant matched the duration of their pauses. That is, before taking a new turn, each partner paused for a duration that roughly matched the other's most recent pause. That allowed them to maintain their "rhythm." The contingent relation between the mother's and the infant's vocalizations prompted some investigators to refer to these vocalizations as "proto-conversations"—an interpretation that seems justified because the infant and the mother alternated their utterances, just as adult speakers and listeners do in real conversations.

Contingent relations between mother and infant are not restricted to vocalization. Many experiments have reported significant correlations between the mother's attentiveness and the infant's smiling and cooing,[50] and between other expressions of mothers' and infants' affect.[51] Developmental psychologists refer to the close temporal relation between infant's and mother's affect and behavior as *dyadic*, to highlight the fact that the coordination of those events contains more information than individual analyses of their behavior.

Dyadic relations are the main bond between an infant and her mother until the infant begins to crawl and explore objects in the environment. Beginning at about six months, *triadic* relations develop among the infant, her mother, and objects of mutual interest. Those relations facilitate the development of joint attention.

Joint Attention: While playing with an infant, it is commonplace for a mother to engage the infant's interest in a toy by looking at the toy. She then waits for the infant to gaze at it and to look back at her and smile. That sequence is an example of joint attention. It provides the first instance in which an infant and another individual share their perception of an environmental event—in this example, knowing that each one saw a particular object.

Joint attention, a nonverbal process, is crucial for word learning. Consider, for example, a mother teaching an infant that the name of the object with which she is playing is *doll*. Without joint attention, *doll* might refer to any other item in the room: a chair, a fan, a shoe, a dog, and so on.[52] However, once the "common ground" of joint attention is achieved, it's easy for the infant to identify the object as a *doll*.[53]

Joint attention is more complicated than shared gaze, a phylogenetically older ability, in which two individuals simply orient toward the same object. To appreciate the difference, imagine that you are walking with a friend and that you turn your head toward a passing car. Your friend does the same. Unless you have some way of communicating what you saw, you have no way of knowing that you both saw the same thing. That's why, in the previous example, smiling after shared gaze is important. It's a nonverbal way of saying "I saw what you saw." For joint attention to signify shared experience, it's important for one person to engage in a communicative act *after* the other person looks at the object in question. Children often do that by smiling, by pointing, or, literally, by offering the object to their caretaker.[54]

Joint attention not only facilitates the acquisition of words but also predicts the size of a child's vocabulary at twenty-four months and older ages. The higher the rate of joint attention at twelve months, the larger a child's subsequent vocabulary.[55] Joint attention is also significant because the words a child learns are declarative and, therefore, part of a conversation. After the mother says *doll*, the child replies *doll* or engages in some other social act to indicate that the child saw it. Such exchanges are the only way infants can begin to learn names.

Intersubjective relations and experience with joint attention provide infants with a crucial foundation for learning to produce their first words. Even as a thought experiment, however, it is inhumane and irresponsible to think about controlling for their influence. There are, however, two "natural" conditions in which that has happened: children raised in orphanages and autistic children. An infant's social relations with her caretakers are compromised in both conditions. The etiology of social impoverishment differs in each instance, but the result is nevertheless the same: language learning is delayed or absent. Both types of problems are discussed in chapter 4.

■ ■ ■

Summary: Charles Darwin's formulation of the theory of evolution focused on variation of anatomical structures and the selection of beneficial changes. Herbert Spencer subsequently argued that natural selection applies to changes in behavior with the same force as it does to structure. That idea gave rise to comparative psychology, a discipline that investigates the behavior of different species. Comparative psychology is relevant to the evolution of language because it asks how learning can account for similarities and differences in human and primate intelligence.

During the twentieth century, behaviorists developed two methods for studying learning: classical and operant conditioning. Classical conditioning is not relevant to language because its focus is involuntary behavior. Operant conditioning, which studies voluntary behavior, has been used to explain a variety of behaviors in animals and humans. Skinner, a major theorist of operant conditioning, argued that it can be used to explain language.

A basic tenet of operant conditioning is that the strength of a response is determined by its consequence: positive or negative reinforcement. Skinner broadened that concept to include an occasioning stimulus. In the presence of that stimulus, a particular contingency is in effect (for example, a green light that tells a motorist that it's appropriate to drive, or a word that conveys a particular meaning, such as common and proper nouns, verbs, and so on). Skinner also proposed a chaining mechanism to create sequences of words that comprise phrases and sentences.

Chomsky criticized Skinner's theory by showing that it could not account for a basic feature of language: the ability to create an infinite number of meanings from a finite vocabulary. Chomsky also showed that the meaning of a sentence was not determined by the relation between successive words, as required by Skinner's theory. In an embedded sentence, for example, such as *The boy who wore a hat smiled*, the relation between *boy* and *smiled* is understood even though those words are not adjacent.

Chomsky concluded that language cannot be explained as conditioned behavior and that it is uniquely human. Some behaviorists nevertheless attempted to train chimpanzees to learn language: specifically, to produce sentences. They used nonvocal languages because chimpanzees cannot articulate the sounds of language. Some projects used American Sign Language. Others used visual languages composed of symbols that differed in size, color, shape, and other details.

Both types of project failed. Although chimpanzees could learn to produce sequences of signs or visual symbols, they never learned the meanings of those signs and symbols. Analyses of chimpanzees' performance showed that the basic problem was not their inability to create sentences but their inability to even learn the meaning of words. Unlike children, who learn that words function as names that can be used to refer to particular objects and events, chimpanzees could only learn the imperative function of signs and visual symbols: a means to obtain a particular reward.

Chomsky's claim that language is uniquely human proved correct. Nevertheless, he was unable to account for the origin of words. Because words evolved before grammar, he could not explain how the gap between animal communication and language was bridged.

The main purpose of this book is to draw attention to the significance of words, and how their origin was the first step toward language. By learning words, some ancestor was the first individual to name and refer to objects and to engage in conversation, all without grammar!

Much has been learned during the past forty years about the origin of words. Developmental psychologists have shown that nonverbal emotional and cognitive relations between an infant and her caretaker are a critical foundation of the infant's ability to produce her first words. The absence of such relations in apes helps to explain why they are unable to learn language.

Paleoanthropologists have discovered fossils of many human ancestors who evolved after hominins split from chimpanzees six to seven million years ago. Those discoveries have not only increased the evolutionary distance between chimpanzees and *Homo sapiens*, but they have also identified a recent ancestor who is a likely candidate for the originator of words.

Chapter Two

APE LANGUAGE

ONE OF Aesop's most famous fables is about a hungry fox who scorned some grapes he couldn't obtain. After giving up, he stalked away, saying, "Those grapes must surely be sour, I wouldn't eat them if they were served to me on a silver platter."

It's not just in Aesop's fables that we find animals who talk. They also appear in nursery rhymes, songs, cartoons, biblical tales, comics, and movies. And it's not just children who believe that animals can talk. So do many readers—you may be one of them. But when pressed, even the most ardent animal lover would agree that a conversation with an animal would be very one-sided.

It is nevertheless not obvious *why* animals can't talk or, conversely, why only humans can. In chapter 1, I proposed a simple answer: All human languages use words. Animal communication does not. Still, that distinction doesn't do justice to the gap between language and animal communication. To say that human languages use words implies a conversation between a speaker and a listener who take turns sharing information.

That may seem obvious, but a moment's thought will reveal that people often regard *any* use of words as linguistic. Consider, for example, a dog that responds appropriately to the verbal commands "sit," "stay," and "fetch ball." Such behavior illustrates what linguists refer to as *comprehension*, but it doesn't qualify as language because it doesn't include the *production* of words. Just because a dog can distinguish the sound "fetch ball" from other sounds doesn't mean that the dog perceives those sounds as words or that it understands their meaning.

Modern linguistic theory, however, which has been dominated by Chomsky's theory of language, regards grammar rather than words as the quintessential feature of language. Because it is assumed that some animals know the meaning of words (as suggested by tests of comprehension), evidence of grammar has been the litmus test of an animal's linguistic ability.

To challenge Chomsky's theory, some psychologists attempted to teach language (that is, its production) to chimpanzees, humans' closest living relative.[1] Others studied gorillas, parrots, dolphins, and sea lions.[2] In the interest of brevity, I focus here on experiments with chimpanzees. However, my conclusions—which the reader should be warned are negative—apply to all claims that animals can learn language.

■ ■ ■

Two approaches have been employed to teach chimpanzees to use language. In "natural" experiments, chimpanzees are raised informally in a home environment, like human children. In laboratory experiments, chimpanzees live in cages and are trained by formal methods that used objective tests to assess knowledge of language.

In a family setting, it was hoped that a chimpanzee would "ape" the language of the family in which it was reared. As with children, it was assumed that a family setting was the best environment for teaching language. Because communication was informal, little effort was made to maintain detailed protocols of the teacher's and the subject's utterances.

In the laboratory, researchers hoped that their results would be easier to replicate than those obtained in home environments. Instead of a learning a natural language, the "lab" chimpanzees were trained to use visual symbols that varied in shape, size, and color. Like words, though, the meaning of those symbols was arbitrary. In one language, for example, a blue plastic triangle meant "apple."[3]

In this chapter, I compare projects that attempted to teach a chimpanzee to use language in a home environment and in a laboratory. Although the form of communication differed in each case, both types of project revealed how difficult it is to avoid anthropomorphizing the behavior of creatures toward whom we are profoundly empathic. My discussion of the intense effort to teach chimpanzees to communicate in sign language or by using visual symbols, the failure of those efforts, and the ease of teaching children to use any language should clarify what makes language special.

LEARNING LANGUAGE IN A HOME ENVIRONMENT

In 1931, Winthrop Kellogg and his wife began the first home-reared ape language experiment in the United States. They succeeded in training Gua, an infant female, to say four words—*mamma, poppa, up,* and *cup*[4]—but even those words were labored and unnatural. Initially, one of Gua's surrogate parents molded her lips and mouth into the correct articulatory position. Gua eventually learned to use her own hands to move her lips and mouth, at which point she exhaled to produce the correct sound. However, she would do that only if she were given a food or a drink reward. Caroline and Keith Hayes performed a similar experiment on a chimpanzee named Vicki and obtained similar results.[5]

Gua's meager achievements were widely interpreted as evidence that chimpanzees lacked the brainpower needed to master language. However, some psychologists argued that the problem was peripheral. They assumed that chimpanzees were intelligent enough to think but that they lacked the articulatory apparatus needed to produce the sounds of spoken languages. Robert Yerkes, a distinguished primatologist, observed that

> great apes have plenty to talk about, but no gift for the use of sounds to represent individual . . . feelings or ideas. Perhaps they can be taught to use their fingers, somewhat as does a deaf and dumb person . . . to acquire a simple, non-vocal language.[6]

PROJECT WASHOE

In 1966, Allen and Beatrice Gardner, of the University of Nevada, began a project to teach Washoe, a one-year-old female chimpanzee, to use American Sign Language (ASL). ASL, which is a natural gestural language used by thousands of hard-of-hearing Americans,[7] was the predominant form of communication in the Gardners' home. Some examples of signs are shown in figure 2.1.

In contrast to the meager results of earlier home-reared experiments, Washoe learned 132 signs of ASL during the three and a half years she spent with the Gardners. She clearly earned a place in history as the first chimpanzee to learn to sign in a natural human language. Of greater significance was her ability to combine signs into sequences that some psychologists interpreted as sentences.

In an early diary report, the Gardners wrote: "Washoe used her signs in 29 different two-sign combinations and 4 different combinations of three signs." That report prompted Roger Brown, the most eminent psycholinguist of the time, to comment, "It was rather as if a seismometer left on the moon had started to tap out 'S-O-S.'"[8]

Brown compared Washoe's sequences of signs to a child's early sentences and noted similarities in their structural meanings (e.g., agent-action, agent-object, action-object, and so on). Those combinations resembled combinations produced by young children. For example, in children, an agent-action construction includes noun-verb sequences such as *Mary play* or

FIGURE 2.1 Examples of signs of American Sign Language (ASL). Like words, most of the signs of ASL are arbitrary. Their configuration does not reveal their meaning.

Photos: Liam Archbold & Natalie Zarrelli.

Daddy kiss. An action-object construction includes verb-noun sequences, such as *throw ball*, or *push truck*, and so on.

To justify those interpretations, linguists typically obtain a "corpus" of a child's utterances: a compendium of combinations of two, three, four or more words and a description of their communicative intent. In a highly influential book, Brown summarized a corpus describing the acquisition of language by three children given the pseudonyms Adam, Eve, and Sarah.[9] Brown's work raises a question as to whether there was a corpus of combinations justifying the claim that Washoe's combinations had structural meanings.

In a diary report, Roger Fouts, Washoe's main trainer, described what is arguably the most famous sequence of signs generated by a chimpanzee: *water bird*. While rowing past a swan, Fouts signed *what's that?*[10] Washoe responded by combining two familiar signs, *water* and *bird*. Even more remarkable was the fact that Washoe had never learned any signs for identifying birds.

Before accepting Fouts's rich interpretation of *water bird*, I think it necessary to eliminate some simpler explanations:

1. Washoe may have been trained to sign *water* and *bird* as separate utterances before she saw the swan.
2. Upon seeing the swan, Washoe may have signed *water* and *bird* as two separate utterances.[11]
3. Washoe may actually have signed *bird water*, but Fouts recorded her utterance as *water bird* because, as an English speaker, he learned to combine adjectives and nouns in that order.

Instead of relying on anecdotal reports, I thought it important to obtain a corpus that documented the chimpanzee's and its trainers' utterances. None existed for Washoe, nor were there any videotapes of Washoe signing with her teachers. I therefore started my own project with an infant male chimpanzee that I named Nim Chimpsky.[12] My goal was to fully document Nim's use of sign language and to obtain a permanent record of Nim's interactions with his trainers.

PROJECT NIM

From the age of two weeks, Nim was raised by a family in New York.[13] As with Washoe, ASL was the major form of communication. Nim's ability to use ASL was documented by his teachers, who whispered the signs he made, and their contexts, into miniature tape recorders. The teachers transcribed their tapes shortly after each session.

By his third birthday, Nim had learned to produce 125 signs. Those are shown in figure 2.2, in the order in which they appeared. Some examples of individual signs that Nim learned are shown in figure 2.3.

What distinguished Project Nim was a corpus of his combinations, which eventually included more than 20,000 combinations of two or more signs. Table 2.1 shows some examples. Many of Nim's combinations appeared to have been generated according to simple grammatical rules, such as *more + x*; *transitive verb + me* or *Nim*. For example, there are 375 instances of *play me* and only 81 instances of *me play*. That type of contrast provided the strongest evidence of an ape's ability to create sentences.

To my utter chagrin, further analysis of Nim's combinations revealed that they were an artifact of imitation and therefore devoid of the meanings I had attributed to them. Ironically, I

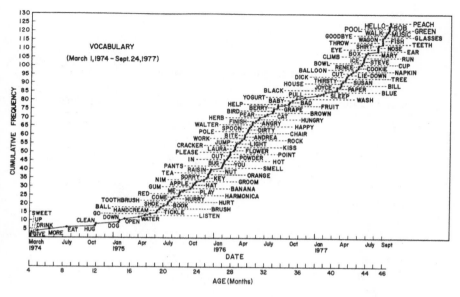

FIGURE 2.2 A cumulative record of the signs Nim learned to produce while he was raised in New York.

Figure courtesy of the author.

apple

red

tickle

cat

FIGURE 2.3 Examples of Nim signing.

Photo courtesy of the author.

TABLE 2.1 Nim's most frequently used two-sign combinations*

Combination	Frequency	Combination	Frequency
play me	375	*tickle Nim*	107
me Nim	328	*hug Nim*	106
tickle me	316	*more drink*	99
eat Nim	302	*eat drink*	98
more eat	287	*banana me*	97
me eat	237	*Nim me*	89
Nim eat	209	*sweet Nim*	85
finish hug	187	*me play*	81
drink Nim	143	*gum eat*	79
more tickle	136	*tea drink*	77
sorry hug	123	*grape eat*	74
		hug me	74
		banana Nim	73

*Terrace, H., Petitto, L., et al. (1979).

made that discovery in 1978 while writing an article in which I argued that Nim had in fact created simple sentences. What made me change my mind was a videotape of Nim signing with one of his teachers—a tape I had previously watched many times. I vividly remember the shock I experienced when I saw Nim's teacher prompt him to sign, less than a fraction of a second before he signed.

Why hadn't I previously seen or noticed his teacher's prompts? The reason was simple. Each time I watched a videotape of Nim signing, I thought I was watching a chimpanzee making history by conversing with a human. It was as if I were observing Nim through a telephoto lens, one that allowed me to zoom in on his signs at the expense of any context. On this occasion, though, my gaze shifted from a telephoto to a wide-angle view. That allowed me to see the relation between the teacher's and Nim's signs.

None of Nim's teachers were aware of their prompting. They told me that Nim's gesturing made it seem as if they were having a spontaneous conversation with him. Videotapes showed otherwise. And it wasn't just members of Project Nim who didn't see the teachers' prompts and their influence. In 1974, a panel from the National Institutes of Health (NIH), including a linguist who was fluent in ASL, watched Nim sign with a teacher through a one-way window of a classroom I had built for him at Columbia University. The panel was so impressed by the quality of Nim's signing that they recommended approval of a grant proposal I had written to fund my research.

Once I saw how Nim's teachers contributed to his signing, I began to document what, to the unaided eye, gave the appearance of actual conversations. Slow-motion analyses of Nim's signing with his teachers, at home and in his Columbia classroom, revealed the following type of exchange. Initially, Nim tried to grab whatever reward the teacher brought with her. Because it was withheld, he occasionally signed all-purpose "wild-card" signs such as *me Nim* or *hug*. On most occasions, his teacher prompted him with appropriate signs, approximately 250 milliseconds before he signed.

Having seen the influence of his teacher's prompts, I reinterpreted sequences of Nim's signs that I had previously thought were sentences. Figure 2.4A shows an example: *me hug cat.*[14]

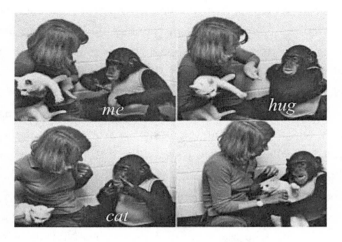

FIGURE 2.4A Nim signing with a teacher. Nim appears to be spontaneously signing the combination *me* → *hug* → *cat.*
Photo courtesy of the author.

A year later, when I next looked at those photographs, the teacher's prompts stared me in the face. Figure 2.4B is identical to figure 2.4A, but in figure 2.4B I have labeled signs that the teacher made just before Nim signed. Before Nim signed *me,* the teacher signed *you.* Before he signed *hug,* she began to sign *Nim,* but Nim crossed her up by making the all-purpose sign, *hug.* Before Nim signed *cat,* she signed *who? Me hug cat* was a sequence of prompted signs, not a spontaneous sentence.[15]

The interaction shown in figure 2.4B is an embarrassing example of what has come to be known as the "Clever Hans" effect—a phenomenon I thought I had learned to avoid as a graduate student. Clever Hans was a German horse who was the subject of much scientific scrutiny at the end of the nineteenth

FIGURE 2.4B Nim signing with a teacher. Signing by Nim's teacher prior to each of the signs shown in the first three panels. Upper left-hand panel: The teacher signed *you* prior to Nim's signing *me*. Upper right-hand panel: The teacher began to sign *Nim* by signing the letter *N*, which he would then use to scratch his forehead (the sign for Nim's name), before Nim signed *hug*. *Hug* is a "wildcard" sign that is often used to obtain a reward. Lower left-hand panel: The teacher signed *who?* before Nim signed *cat*. Nim was "rewarded" with the cat only after he signed *me hug cat*.

Photo courtesy of the author.

century.[16] His trainer claimed that Hans could solve arithmetic problems. For example, when the trainer wrote "3 × 2 = ?" on a blackboard, Hans tapped his foot six times.

It took a distinguished group of scientists more than a year to discover how Hans solved that and similar problems. After posing a problem, the trainer held his breath, concerned about Hans's ability to respond correctly. Hans started tapping when the trainer inhaled. After he tapped the requisite number of times, the trainer exhaled, relieved that Hans could solve the problem. At that point, Hans stopped tapping and was given a reward, usually an apple.

The moral of this tale is to be wary of transmitting unconscious cues when training an animal. I remembered that stricture while performing dozens of experiments on pigeons, rats, and monkeys. In each instance, I avoided making a Clever Hans mistake by using an automated apparatus that prevents such cues. Twenty-five years after I obtained my PhD, I lowered my guard and fell victim to the Clever Hans effect, on the first occasion in which human intervention was needed to train an animal.

Slow-motion analyses of films of other signing chimpanzees also revealed the influence of a teacher's prompting. One example, from a commercially available film, shows Washoe signing *me eat time eat*.[17] In a frame-by-frame analysis, shown in figure 2.5, Beatrice Gardner, Washoe's trainer, uttered three of those signs just before Washoe did. Another film, not reproduced here, shows Koko, a gorilla, signing with her trainer (Penny Patterson).[18] Patterson's cuing is blatant enough for even a naïve observer to see while watching that film in real time.

Nim's corpus, the only one available of a chimpanzee who had been taught to sign, provided an independent opportunity to assess his grammatical ability. In 2013, Yang analyzed Nim's use of signs to determine if his combinations were constructed the way children construct phrases.[19] As an example, consider two-word noun phrases, which often consist of a function word, like *the* or *a*, and a content word, such as *cookie* or *doggie*. If children memorized noun phrases, particular function and content words would

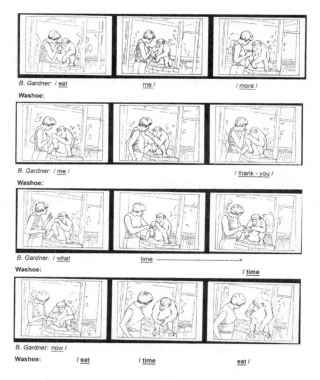

FIGURE 2.5 Tracings from a film of Washoe signing with B. Gardner. Adapted from R. A. Gardner & B. T. Gardner (1973). An earlier version appeared in Terrace, Petitto, et al. (1979). Figure courtesy of the author.

occur together: for example, *the cookie* or *a cookie*, at the expense of *a cookie* or *the cookie*, or *the doggie* at the expense of *a doggie*, and so on. Such repetitions would yield a low diversity score. If, in contrast, they selected function and content words independently, which is what a grammatical rule would predict, diversity would be high. That's because the choices for generating a noun phrase would range over all of the words a child knows.

After analyzing thousands of noun phrases of children, Yang showed that their diversity was high, suggesting that children combined function and content words independently. That was not true of Nim's combinations. Nim's eight most frequent two-sign constructions were: (*more/give*) X, X (*more/give*), verb (*me/Nim*), (*me/Nim*) verb, food-item (*me/Nim*), (*me/Nim*) food-item, nonfood-item (*me/Nim*) and (*me/Nim*) nonfood-item. Yang's analysis of these eight two-sign constructions produced low diversity scores. That would be expected if Nim's combinations were not generated by his following a grammatical rule. A low diversity score would also be expected if Nim's combinations were imitative of his teacher's signs.

CRITICISMS OF PROJECT NIM
AND A WAY TO RESOLVE THEM

My conclusions about Project Nim have been criticized on various methodological grounds by other investigators of ape language projects.[20] Those investigators have not, however, revealed enough information about their procedures to show that they were immune to the same criticism.

The Gardners, for example, argued that Nim had too many teachers. In my book about Nim,[21] I listed the names of fifty-eight volunteers who interacted with Nim while he lived in New York. Most of those volunteers spent only a few weeks with Nim. They didn't qualify as regular teachers because they failed to inspire enough trust. During the four-and-a-half-year period in which Nim was in New York, he had only eight regular teachers, each of whom spent many hours a week working with him. I listed the names of volunteers to thank them for their efforts.

Project Nim was run as an extended family in which many members were temporary babysitters. Given the need to cover all of Nim's waking hours, the size of our group was appropriate: a core group supplemented by a few volunteers. The Gardners never provided a complete list of the members of their project. I would be surprised if its size differed from that of Project Nim.

The Gardners also argued that Nim's signing was motivated by primary rewards. But, as far as I'm aware, *all* ape language projects used primary rewards to motivate signing. I noticed that practice while visiting various other ape language projects, including the Gardners'. In each instance, it was obvious that signing would stop if the chimpanzee didn't receive occasional "freebies."

These and other criticisms of Project Nim should not detract from its major conclusions. Chimpanzees trained to use sign language do so only if they are prompted by a human trainer. Even then, they will only sign to obtain a reward. Our discourse analysis showed that a chimpanzee's signing differs fundamentally from the spontaneous discourse of infants and children. Yang's analysis produced a similar conclusion.

It is, of course, possible that my conclusions about a chimpanzee's inability to learn language are wrong. It has been suggested, for example, that a project in which a chimpanzee was trained by a smaller group of fluent signers would yield more positive results.

A few years after publishing my results, I stated that "my conclusions could be disproved by an unedited videotape in which the chimpanzee and its trainer were visible in each frame."[22] That tape would reveal the extent to which the trainer prompted the chimp and whether it was given small rewards to motivate it to sign. Videotaping is inexpensive and requires little effort. More than thirty years later, I have yet to see any evidence that challenges my conclusions. Given the significance of contrary evidence and the ease of making a videotape, its absence speaks volumes.

Project Nim revealed how easy it is to read into a chimpanzee's gestures meanings that are entirely in the eye of the beholder. Part of the problem is the likelihood of a Clever Hans effect. That possibility can be eliminated by training language with a computer. In the following section, I describe two projects, one of which actually trained chimpanzees to respond to visual symbols that appeared on a computer keyboard, thereby eliminating the possibility of Clever Hans-type cues.

LEARNING LANGUAGE IN A LABORATORY

PREMACK

Some psychologists opted for a laboratory setting for training a chimpanzee to use language because such a setting would provide more objective records of their subjects' achievements. Instead of using a gestural language, they invented languages in which the "words" were geometric symbols. The first such experiment was performed by David Premack of the University of Pennsylvania.[23]

Premack used plastic chips of different shapes, sizes, and colors as symbols for words. Each symbol had a magnetic backing that allowed it to be secured to a "language board." Examples are shown in figure 2.6A[24] and 2.6B.[25] Premack's protocol for teaching the meaning of each symbol was a model of simplicity. To train Sarah (his best student), he first gave her a few pieces of apple. He then placed a blue plastic triangle, the symbol for *apple*, within easy reach. Sarah had to place it on a vertical language board to earn a piece of apple.

Symbols that weren't nouns were trained in a similar manner. Consider how Sarah learned to use the symbols *same* and *different*. On each trial, those symbols were placed in front of her along with two objects: for example, fruits, bowls, dishes,

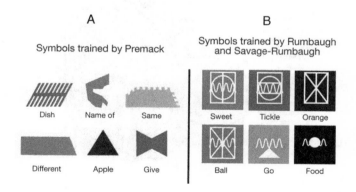

FIGURE 2.6 Examples of lexigrams used in projects by (*a*) Premack and (*b*) Duane and Sue Savage-Rumbaugh.

and so on. When both objects were physically the same (say, two bananas), Sarah was rewarded for placing the symbol *same* between them. When they were different (say, a banana and an orange), Sarah was rewarded for using the symbol *different*.[26] The symbols for *same* and *different* are shown in the left-hand portion of figure 2.6A.

Once Sarah learned a small vocabulary of plastic symbols, she was trained to produce a sequence of symbols. If, for example, her teacher wanted her to produce the sequence *Mary give apple Sarah*, Sarah had to place those symbols on the language board, in that order. Figure 2.7 shows a photograph of her receiving a reward after producing such a sequence.

FIGURE 2.7 Training Sarah to use symbols by placing them on a metallic language board. The language board provided a visual record of Sarah's use of symbols.

Photo from D. Premack (1976). Reprinted by permission of *American Scientist*, magazine of Sigma Xi, The Scientific Research Society.

Sequences were trained in steps. Sarah was first rewarded for combining *give* and *apple* in that order, but not *apple give*. Sarah was then taught to produce the three-item sequence *Mary give apple*. To help them learn the names of other chimpanzees, each chimpanzee wore a necklace with a distinctive plastic chip. How did Sarah learn her name? When another chimp was present during training (say, Gussie), the other chimp was given the apple if Sarah produced the sequence *Mary give apple Gussie*. Following that experience, Sarah rarely made the same mistake.

SAVAGE-RUMBAUGH AND RUMBAUGH

In 1973, Duane and Sue Savage-Rumbaugh, of the Georgia State Language Research Center, automated Premack's method by training chimpanzees to touch symbols on a computer keyboard.[27] They called that language *Yerkish*. It consisted of "lexigrams which were unique combinations of particular colors and geometric configurations." *Food*, for example, was defined as a sine curve, on which a white dot was superimposed, on a black background. Examples of Yerkish lexigrams are shown in figure 2.6B. To obtain a reward, Lana had to touch the appropriate lexigram. Touching any of the other lexigrams produced either no reward or one that was less desirable. Because Yerkish was administered by a computer, it eliminated the possibility of a Clever Hans effect. A photograph of Lana, one of many chimpanzees trained by Rumbaugh and Savage-Rumbaugh, is shown in figure 2.8.

The procedure used to train sequences was similar to Premack's. If, for example, Lana wanted a piece of apple, she had

FIGURE 2.8 Lana using Yerkish lexigrams to communicate her needs.
Photo from Rumbaugh (2013). Reprinted with permission.

to enter the sequence *please machine give apple*. Lana would not receive a reward if she entered *please machine apple give*, or *machine give please apple*. Other rewards, such as an opportunity to look out the window or to look at pictures of familiar objects, required slightly different sequences, such as *please machine open window; please machine show slides.*

Lana learned to produce sequences of lexigrams for the same reason that Nim learned to produce sequences of signs. Both chimpanzees were motivated to "speak" by reward rather than by having a conversation with their trainers. Neither type of sequence was grammatical. In Nim's case, that was evident in videotapes of his interactions with his teachers. For Lana, it was evident in an analysis of a corpus of her utterances in Yerkish.

In 1980, Thompson and Church analyzed a corpus containing more than 14,000 sequences that Lana produced during a three-year period. They showed that most of her combinations were the result of two nongrammatical rules that Lana used on each trial.[28] First, she chose one of six stock sequences: for example, *please machine give, please machine make, please put in machine,* and so on. Lana then added a lexigram for a particular reward, such as *apple, banana, orange,* and so on.

Those rules accounted for 91 percent of Lana's combinations. If, for example, a reward was in view (say, an apple, chocolate, banana, or the like), Lana would have to produce the stock sequence *please machine give X*, where X was the reward. If it were not in view, she would have to produce the sequence *please machine put into machine X*. For a nonedible reward (say, music or slides), she would have to produce a stock sequence such as *please machine make X*.

Thompson's and Church's analysis appeared to have a sobering effect. Instead of trying to show that chimpanzees could create a sentence, Savage-Rumbaugh and Rumbaugh turned to experiments that investigated a chimpanzee's understanding of individual lexigrams. Two such experiments struck me as being the best of their kind. One attempted to show that a chimpanzee could use a lexigram to communicate with another chimpanzee, albeit only to obtain a primary reward. The other attempted to show that a chimpanzee could use a lexigram to categorize other lexigrams. Were it not for a confound, which I describe later in this chapter, it would have shown, for the first time, that a nonhuman primate could use lexigrams to think about other lexigrams.

Communication in Yerkish

To show that chimpanzees can communicate with lexigrams, Savage-Rumbaugh and Rumbaugh trained Sherman and Austin to use tools to obtain food or drink rewards that were otherwise inaccessible.[29] Specifically, Sherman and Austin learned how to use a key to unlock the padlock of a box, money to operate a vending machine, a straw to drink from a long container, a stick to push food out of a hollow tube, a sponge to soak up liquids from a deep container, and a wrench to remove the bolts that secured a lid to a chest. After learning to use those tools, Sherman and Austin were trained to request a particular tool by touching the relevant lexigram on their computer consoles. Sherman and Austin were also trained to share the food they obtained by using tools.

That set the stage for a paradigm in which Sherman and Austin could request a tool from each other to help them retrieve an otherwise inaccessible reward. Each chimpanzee was housed in a separate room, in which it could watch the other through a glass window. One room had a container from which food or drink could be obtained only by using a tool. The other room had a box that contained the six tools the chimps had been trained to use. An opening under the window allowed one chimpanzee to pass a tool through the wall separating their rooms.

Because Sherman and Austin took turns occupying each room, I will refer to them as Chimpanzee 1 and Chimpanzee 2. Prior to the start of each trial, an experimenter baited a container with food or drink, in full view of both chimpanzees, and then left the room. Chimpanzee 1 could request the tool needed to provide access to the inaccessible reward by touching the relevant lexigram on his console. That lexigram was then illuminated in Chimpanzee 2's room. After Chimpanzee 2 retrieved the correct tool and passed it through the opening between the two rooms, the experimenter allowed Chimpanzee 2 to enter Chimpanzee 1's room, and encouraged them to share the reward.

On most trials, the outcome was successful. One chimpanzee selected the appropriate tool and passed it to the other chimpanzee. Of particular interest was the behavior of Chimpanzee 1 when Chimpanzee 2 selected the wrong tool. What followed is the only instance of which I'm aware in which one chimpanzee attempted to influence another chimpanzee's behavior by pointing. When the wrong tool was selected, Chimpanzee 1 often banged on the window and pointed to the toolbox in Chimpanzee 2's room, as if to say "not that tool, the other one!"

Much training was needed to establish this kind of communication, and it only occurred when Sherman and Austin were motivated to obtain rewards. It is nevertheless a remarkable achievement because such performance has never been observed in the wild.

"Thinking" in Yerkish

To show that a chimpanzee could think about a lexigram, Savage-Rumbaugh and Rumbaugh designed a task in which Sherman and Austin had to use the lexigrams *food* and *tool* to categorize other lexigrams.[30] First, Sherman and Austin had to sort, into separate bins, three exemplars of food (orange, bean curd, and bread) and three exemplars of tools (a key, money, and a stick). They were then trained to label each of those objects with the lexigram *food* or *tool*.

Sherman and Austin could perform that task either by forming an association between each item and the lexigrams *food* or *tool*, or by using the lexigram *food* for edible items and the lexigram *tool* for inedible items. Savage-Rumbaugh and Rumbaugh sought to demonstrate the latter possibility by seeing if *food* and *tool* generalized to lexigrams that Sherman and Austin had not physically sorted. Simply put, they were asking whether a chimpanzee could think of a banana as an item of food just by seeing the lexigram *banana*, or a wrench as a tool just by seeing the lexigram *wrench*.

For that purpose, they tested Sherman and Austin on the chimps' ability to categorize twenty-eight previously learned lexigrams (fourteen foods and fourteen tools), none of which had previously been associated with the lexigrams *food* or *tool*. During the test for categorization, Sherman responded correctly on all of the trials; Austin, on all but one.

Unfortunately, it doesn't follow that Sherman and Austin were thinking about food or a tool when they used the lexigrams *food* or *tool* to respond to the novel lexigrams. Logically, that task was simply a choice between a food and a nonfood or between a tool and a nontool.[31] Sherman and Austin only had to think of one category, or its absence—a so-called default strategy—to perform correctly.

To eliminate the possibility of using a default strategy, it is necessary to train three categories: say, flowers, toys, and tools.

In that case, the subject would have to represent two categories when making a choice. Suppose, for example, that the subject was shown the lexigram *orchid*. To respond correctly, it would have to choose between the representations *flower* and *toy*, or between the representations *flower* and *tool*. With three categories, it isn't possible to use a default strategy. Accurate performance with three choices eliminates the confound inherent in the binary choice paradigm, which I described earlier.

Instead of pursuing research on the functions of individual lexigrams, Savage-Rumbaugh returned to the question of grammar: specifically, to what extent could a chimpanzee understand a sentence? For that, she turned to another species.

Kanzi, the Most Linguistically Advanced Chimpanzee?

Lana, Sherman, Austin, and more than a dozen other chimpanzees who learned Yerkish were common chimpanzees (*Pan troglodytes*). To study grammatical ability, Savage-Rumbaugh focused on bonobos (*Pan paniscus*), a more recently evolved species, whose members are smaller, less aggressive, and more playful. Among nonhuman primates, bonobos also manifest an unusually high degree of interindividual tolerance. Mothers permit other adults, both male and female, to carry and play with infants, and males frequently share food with infants.

Kanzi, the first bonobo Savage-Rumbaugh trained, often accompanied his foster mother (Matata) while she was taught Yerkish. Kanzi, however, was not given any formal training. Much to Savage-Rumbaugh's surprise, Kanzi not only learned to master many lexigrams that were difficult for Matata to learn, but also hundreds more.[32] Kanzi also learned to respond to many spoken English commands (e.g., *Go to x, Give x to y,* etc.). While giving such commands, the trainer often pointed to relevant objects, places, or individuals.

Kanzi was reported to have learned spoken English "spontaneously," without explicit training, presumably as a young child does. Although Kanzi used lexigrams to communicate his desires to his trainers, they also spoke to him in ordinary English. In this manner Kanzi supposedly learned more words in English than in Yerkish.[33]

Kanzi is best known for his performance on tests of comprehension of spoken English. In one experiment, Savage-Rumbaugh compared Kanzi's performance with that of a two-year-old child. Kanzi was eight years old at that time. Both Kanzi and the child were given 660 instructions in spoken English, such as *Pour the milk on the vacuum,* or *Go get some cereal and give it to Rose.*[34] Savage-Rumbaugh assumed that correct responses to those commands required an understanding of their grammatical structure.

Part of the motivation for this study was the criticism that the sequences produced in experiments using sign language were not grammatical because they were prompted and/or imitative of a teacher's signs. Savage-Rumbaugh also argued that, in all spoken languages, comprehension normally precedes production. On that basis, she argued that comprehension was evidence of linguistic knowledge.

Kanzi's ability to comprehend spoken English was tested with a double-blind procedure. The experimenter and Kanzi were in separate rooms. The experimenter observed Kanzi's behavior through a one-way mirror. That prevented the experimenter from conveying any nonverbal cues, such as looking at, or pointing to, objects. An accomplice who could not hear the command observed Kanzi's behavior following each command: *hide the toy gorilla, now go get your ball*, and so on.

All of the commands on which Kanzi was tested were novel. Although he had learned to differentiate *milk* and *cereal* from other foods, *vacuum* from other objects, and *Rose* from other individuals, he was never tested on commands that required him to understand the relation between two words.

Kanzi's performance was considered correct on a majority of trials. Because his accuracy was slightly higher than the child's (72 percent versus 66 percent), Savage-Rumbaugh argued that he understood English and that he "clearly processed semantic and syntactic features of each novel utterance."[35]

There are many reasons to question that claim. Here I only raise some methodological questions and note that similar performance was obtained with dogs, a species not known for its linguistic ability. To respond correctly, Kanzi only had to interpret two words. For example, to respond correctly to *Pour the milk on the vacuum*, it is not necessary to understand *pour*, *the*, or *on*. Also, from previous training, Kanzi knew that vacuums couldn't be poured. Thus, there is no reason to expect his performance to differ if the command was *Pour the vacuum on the milk*.

Kanzi's performance also received credit for partially correct responses. In the previous example, he would have gotten credit for pouring milk on the floor. Savage-Rumbaugh did not provide enough documentation of her grading system to explain when and how partial credit was given, but we know it was awarded on many trials. For example, Kanzi received partial credit for eating a piece of pineapple in response to the command *Hide the pineapple*,[36] and for breaking off some pine needles in response to the command *Cut the pine needle with the scissors.*[37]

To test Kanzi's understanding of grammar, it is necessary to show different responses to commands such as "Man bites dog" and "Dog bites man." Only 21 of the 660 commands were given in that format. An analysis of Kanzi's performance showed that he responded accurately to only twelve such pairs, a level of accuracy that is close to chance.[38] Even that estimate is generous because of partial credit. With toys, for example, he was asked to *Make the doggie bite the snake* and *Make the snake bite the doggie.* Kanzi put the snake in the dog's mouth in both cases. Both responses were nevertheless given full credit. When such false positives are excluded, Kanzi's success rate was only 30 percent.

No evidence was provided to justify Savage-Rumbaugh's claim that Kanzi learned English without any specific instructions. That assertion would be defensible if Kanzi's training had not included any *formal* reward contingencies, but there were clearly many *informal* contingencies. For example, Savage-Rumbaugh noted that "[f]ood was dispersed at identifiable locations on a 55-acre wooded area, and most of the day was spent traveling from one food source to another."[39]

Unfortunately, Kanzi's linguistic achievements have only been described anecdotally. That problem could be eliminated by describing his performance in peer-reviewed journals that analyzed publicly available videotapes.

In the introduction to a book about Kanzi, Savage-Rumbaugh commented:

> When I observe a bonobo, it is as though I am standing at the precipice of the human soul, peering deep into some distant part of myself. This is a perception I cannot shake off or dissuade myself from, no matter how often I try to tell myself that *I have no definitive scientific basis for these impressions* [italics added].[40]

Savage-Rumbaugh's empathy with the intelligent bonobo is endearing, but its anthropomorphic quality makes her claims about Kanzi scientifically suspect.

Word Comprehension by Dogs?

About ten years after Savage-Rumbaugh tested Kanzi's understanding of spoken English, three experiments on dogs were published that describe similar behavior. The first experiment showed that a border collie (named Rico) could fetch objects such as balls, Frisbees, toys, and so on following a verbal command.[41] Rico was trained from the age of ten months to fetch objects in different locations in his owner's home and was rewarded with food or play each time he retrieved the correct item. Once that behavior was established, new objects were gradually added. With the dog's owner waiting in a separate room, Rico retrieved the correct object on thirty-seven out of forty trials. That arrangement ruled out a Clever Hans explanation of his performance. All told, Rico learned the names of more than 200 objects.

Of special interest was Rico's ability to learn the names of new items by the process of exclusion. In children, that phenomenon is called *fast mapping*.[42] Fast mapping in dogs was shown in a procedure in which a novel item was placed with a large group of familiar items. On the first trial, Rico was asked to bring one of the familiar items. On the second trial, he was asked to bring the new item. Rico was successful on more than 70 percent of such trials, even though he had never previously retrieved any of the objects he was asked to fetch. He was able to link the novel word to the novel item either because he knew that the familiar items already had names or because they were not novel.

This experiment shows that Rico reliably associated arbitrary acoustic patterns (human words) with specific items in his environment. A similar experiment was performed with Chaser, another border collie, who was able to fetch 1,022 different objects.[43] Because it was hard for an experimenter to remember the names of that many objects, their names were inscribed on each object with an indelible marker: for example, *elephant, lion, seal, tennis, choo choo, mickey mouse, bling, nosey, Santa Claus*, and so on.[44]

A third experiment showed that Chaser could follow spoken commands to perform particular operations on particular objects (e.g., pawing, nosing, and fetching an object). Chaser was trained to combine the three verbs *nose, paw*, and *fetch* with one of the nouns of her vocabulary: for example, *nose lamb* or *paw ABC* (an *ABC* was a cloth alphabet block).

Neither experiment claimed that border collies have any linguistic knowledge. What they did show is that a highly socialized animal has the ability to differentiate the sound patterns of English words and to use them as discriminative stimuli to solve a problem through which the animal could earn food or an opportunity to play. These experiments have an important implication: It is misleading to evaluate linguistic knowledge with tests of comprehension. Tests of production provide much stronger evidence.

THE MEANING OF SYMBOLS IN "APE LANGUAGE" EXPERIMENTS

Training an animal to produce arbitrary sequences is no guarantee that those sequences are meaningful. This can be seen when comparing the sequences of lexigrams produced by Lana and other chimpanzees with those of monkeys who learned to produce similar sequences in experiments that made no claims regarding linguistic knowledge.

Superficially, the sequences of symbols produced by Lana and other chimpanzees seem to be sentences. More careful analyses, however, showed that the chimpanzees that produced them did not understand their meaning. When a child requests an apple by saying "Please, Mom, pass the apple," it's reasonable to assume that the child knows the meaning of each of those words. When Lana produced the sequence *please → machine → give → apple*, there is no reason to believe that she knew the meaning of *please, machine,* or *give*.[45]

Was Lana being polite to the computer? Of course not. She learned to touch *please* simply to obtain a slice of apple.

Although there is evidence that Lana knew the difference between the meanings of *apple* and other symbols, there is no evidence that she knew the meaning of *machine* or *give*. That would require the ability to contrast *machine* with other nouns or *give* with other verbs.

The folly of interpreting *please* → *machine* → *give* → *apple* and similar sequences as sentences is also illustrated by a project that I began in the 1990s, in which I showed that rhesus monkeys could learn rote sequences composed of photographs of natural objects.[46] The monkey had to respond, in a particular order, to seven photographs, all of which appeared simultaneously on a touch-sensitive video monitor: mountain → large birds → frog → deer → man → dolphin → foxgloves. These sequences are as meaningless as the rote sequences one uses to obtain cash at an ATM.

Representative trials in which a monkey was trained to produce an arbitrary sequence are shown in figure 2.9. On each trial, the physical configuration of the list items is varied randomly. That prevented the monkey from using an item's location as a cue for selecting the next item (as can happen, for example, when making a telephone call). The change in the configuration of list items from trial to trial made this problem harder than entering a sequence on a number pad at an ATM or making a phone call. In those instances, the position of the numbers remains constant from trial to trial. Instead, the monkey had to remember each item's ordinal position in the sequence without the help of an external cue (e.g., the position of the first item, the second item, etc.). Errors or incomplete sequences resulted in no reward, a time-out, and a new trial. An example of a sequence in which the monkey made an error is shown in the top portion of figure 2.9. The monkey earned a banana pellet if, and only if, it responded in the correct sequence to *all* of the items. That is shown in the bottom portion of figure 2.9.

In many ways, the lists on which the monkeys were trained were more difficult than those learned by chimpanzees. For monkeys, changing the physical position of the list items from trial to trial made the problem more difficult than the one

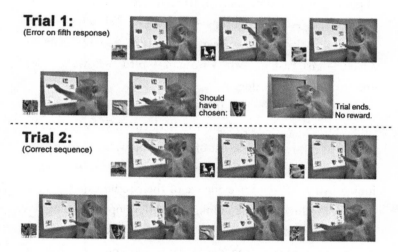

FIGURE 2.9 Photographs of a monkey responding on a seven-item successive chain during two trials (subject made an error while responding to the fifth item in the first trial). The configuration of photographs is changed after each trial. A reward (a banana pellet) was provided after the subject responded correctly to *all* of the photographs, A → B → C → D → E → F → G (A = mountain, B = large birds, C = frog, D = deer, E = a man, F = dolphin, and G = foxgloves).

Photos courtesy of the author.

posed for chimpanzees because the chimpanzees responded on a keyboard on which the position of the lexigrams never changed. Training monkeys to learn seven-item, rather than four-item, sequences not only increased the task difficulty but also posed a problem of interpretation. The folly of assigning linguistic interpretations to the sequences produced by chimpanzees can be seen in interpreting the longer sequences produced by monkeys. Adding *been* → *good* → *today* to the sequence *please* → *machine* → *give* → *pellet* is foolish enough. The reader is challenged to interpret the other three sequences of photographs that monkeys learned to execute: List 2: hog → grass → baby → mountaintops → parrot → bagel → fish. List 3: cactus → rocky shore → bear → cormorant → butterfly → flower → girl. List 4: moth → woman → bird → flowers → sea urchins → cactus → dog.

As with chimpanzees, it took a considerable amount of time to train monkeys to learn rote sequences. They required a month to master the first list of photographs, but less time to master subsequent lists. Only a week was needed to learn their fourth list. That trend is evidence of their serial expertise, but it doesn't follow that the monkeys had developed any linguistic ability. If we used the logic of Lana's trainers to interpret the meaning of their sequences, we would conclude that the monkey was saying, *please* → *machine* → *give* → *pellet* → *been* → *good* → *today*.[47]

Learning a rote sequence raises an intriguing question: What meaning(s) can be assigned to the items that comprise the sequence? They are clearly not words, at least not in the normal sense. But without meaning, a string of symbols is not a sentence. To see why, let's return to the sequence *please machine give apple*. If required by the experimenter, Lana could just as well have learned *apple give machine please, give please apple machine, give apple please machine,* or the like.

The sequence of signs *me hug cat* (cf. figure 2.4A) raises a similar question. When Nim was taught to sign *cat,* Nim tried to grab the cat from the carrying box in which it was brought to the classroom. After his teacher prevented that from happening, Nim allowed his teacher to mold his hands to sign *cat* (cf. figure 2.3.). That process had to be repeated each time the cat was brought to the classroom.

On many occasions, Nim's teachers tried to get Nim to use other signs when he wanted to play with the cat: for example, *me* and *hug*. With the help of those prompts, Nim began to sign *me hug cat* or *Nim hug cat* more often than not, but in each instance his only goal was an opportunity to play with the cat. Were it not for the teacher's prompts, and her modeling of *me hug cat,* Nim could just as well have produced the sequences *cat me hug, hug cat me,* or something else entirely.

ARE IMPERATIVES WORDS?

Although linguists classify imperatives as words, they do so only to distinguish them from declaratives, which are a bidirectional form of communication between a speaker and a listener.[48] For chimpanzees, imperatives are unidirectional in the sense that they do not require a verbal response.

As we saw in the case of chimpanzees trained to respond with Yerkish symbols, rewards were delivered impersonally by a computer, just as they would be if the chimpanzee were operating a vending machine. Even though Premack taught his chimpanzees to manipulate visual symbols manually to obtain rewards, there is no reason to believe that the outcome would have been different if they had been trained by a computer.

The same is true of chimpanzees that were taught to use sign language. Their trainers had the impression that they were having conversations when they signed *good*, *right*, and so on, after a chimpanzee made the appropriate sign(s), but the only motivation for the chimpanzee to sign was to obtain a reward. Project Nim showed that, despite the best intentions, the only role of the trainer was to prompt signing and not to engage the chimpanzee in conversation. For that reason, it's easy to imagine a chimpanzee learning to sign if he was engaged by a robot that could prompt and detect particular signs.

In all human languages, imperatives constitute a minuscule fraction of a speaker's vocabulary. They differ fundamentally from declaratives because declaratives require a social response. When using declaratives, children share information with other individuals, typically the child's caretaker. In that sense, declaratives are bidirectional. As we will see in chapter 4, a child who learns to say *doggie* or *tree* expects social acknowledgment from the person she addresses, rather than possession of a dog or a tree. Social acknowledgment can take many forms: shared gaze, smile, pointing to the object about which the child comments, or a verbal comment such as *Oh, that Fred's dog* or *Yes, that's a big tree*.

What's missing from all of the ape language projects is conversation. Although chimpanzees can learn imperatives, so can dogs, monkeys, and other species. What distinguishes language is the ability to use words declaratively: that is, conversationally. Until that level of communication is achieved, it is premature to ask if another species can learn a grammar. Instead of asking whether an ape can create a sentence, the question should have been "Can an ape learn to use words?" In both instances, the answer is no.

■ ■ ■

Summary: One of the oldest questions of comparative psychology is whether an animal can acquire language. During the second half of the twentieth century, some behaviorists attempted to train chimpanzees to learn simple features of language, in particular the production of sentences. Those projects used ingenious methods to get around chimpanzees' vocal limitations. Some taught chimpanzees to use the signs of American Sign Language. Others used artificial visual languages in which symbols were distinguished by their size, shape, color, and other physical features.

Two of the projects that used ASL yielded evidence that chimpanzees could combine individual signs to produce phrases, such as *water bird* and *me hug cat* (Project Washoe and Project Nim, respectively). Subsequent analyses showed that such sequences were not grammatical. Unlike human children, who use words to name things in conversational exchanges, chimpanzees only used signs to request or earn rewards. Slow-motion analyses of videotapes and films confirmed that most of the subject chimpanzees' utterances were prompted by their teachers.

Projects that used visual languages succeeded in training chimpanzees to produce sequences of symbols, such as *please machine give apple*. Superficially, those sequences seem like sentences. However, unlike sentences, which occur spontaneously, it took hundreds of trials to train the apes on those sequences, and the sequences had to be learned by rote. With the exception of symbols that referred to particular rewards, there was no evidence that chimpanzees understood their meaning. Analyses of a corpus of sequences of visual symbols showed that they could have been produced by the application of two nongrammatical rules: the combination of a stock sequence of meaningless symbols and a symbol for a particular reward. As with sequences produced in sign language, their only function was to obtain a reward for the subject.

With one exception, all of the projects that attempted to train chimpanzees to learn a language studied the chimps' production. The exception was a project that trained a chimpanzee to respond to spoken commands in English. Grammatical knowledge was not needed to explain performance in response to such commands: the chimpanzee only had to discriminate two words of those sequences.

Experiments with other species showed how easy it is to train superficial features of language. Dogs can respond to spoken commands and monkeys can readily learn to produce rote sequences. Given the complexity of language, such performance is not surprising. However, if our goal is to understand the evolution of language, our focus should be on the steps that were necessary to shift from animal communication to the production of words.

Ape language projects have shown that chimpanzees can learn the imperative function of symbols (that is, how to use them to obtain rewards). Although children learn to use words as imperatives, imperatives are only a tiny portion of their vocabulary. Language would never develop if children were limited to learning imperatives. In that sense, the failure of these projects can be attributed to an ape's inability to learn that things have names and that words can be used conversationally.

Chapter Three

RECENT HUMAN ANCESTORS AND
THE POSSIBLE ORIGIN OF WORDS

NOW THAT we know that chimpanzees, our closest living ancestors, are unable to learn language, we are left with the basic
problem of language evolution. Which, if any, of our not currently living ancestors were the first to use language, why might
they have done so, and what might they have said? Recent discoveries about ancestors who evolved after chimpanzees provide some tentative answers. Before considering them, we must
understand how those ancestors differed from chimpanzees and
the challenges they faced for survival.

The source of such knowledge is *paleoanthropology*, the scientific study of species that evolved after human ancestors and
chimpanzees diverged some 6 to 7 million years ago. With their
discoveries of more than two dozen species that filled that gap,
paleoanthropologists have, for the first time, provided a firm
foundation for theories of human evolution. In so doing, they
confirmed Darwin's hypothesis that humans evolved from apes

in Africa (as opposed to Europe or Asia). However, they also refuted his hypothesis that walking upright and a large brain evolved at the same time. Upright posture developed long before the brain grew to its present size.

Equally important is evidence that challenges Darwin's linear view of human evolution. Darwin assumed that humans and great apes shared a common ancestor from which the human line branched off. That branch was occupied by a handful of species, each evolving from the previous one. By providing evidence that some of our ancestors lived at the same time, paleoanthropologists showed that they could not have evolved in a linear progression. It is more accurate to say that they were related to each other as branches are related to a bush.

Before paleoanthropologists discovered how many human ancestors evolved after we split from chimpanzees, the distance between *Homo sapiens* and chimpanzees seemed much smaller than it does today. Now that we know more about our ancestors, it seems as foolhardy to try to teach language to a chimpanzee as it would be for a pre-Columbian navigator to think that crossing the ocean (inventing words?) would be a simple matter, let alone circumnavigating the world (creating sentences?).

Just how, one might ask, are discoveries about recent ancestors relevant to the evolution of language? The answer lies in imagining the conditions under which one of our ancestors stumbled upon language a few million years ago. Paleoanthropologists have provided much information on that topic, some of which appears to be solid enough to suggest how the use of words contributed to the survival of one of those ancestors.

Even if that hypothesis is correct, I should hasten to add that knowing how words originated would tell us nothing about the origin of grammar. A theory about the origin of words would, however, for the first time describe the appearance of what was the first step in differentiating animal communication from language.

This approach is the opposite of Chomsky's. Chomsky speculated that grammar resulted from a recent mutation, but wrote off the origin of words as a "mystery."[1] Even if Chomsky's hypothesis that a mutation gave rise to grammar is correct, it would not explain how language evolved, because the use of grammar presupposes ancestors who had a vocabulary. That is why this chapter focuses on the origin of words and the circumstances that led to their occurrence.

■ ■ ■

It's an understatement to say that the job of a paleoanthropologist is like looking for a needle in a haystack. It's actually worse: The haystack extends over entire continents, with no clues about the location or shape of the "needle" (fossils). Finding a complete skeleton is very rare. Many human ancestors met death in the jaws of a predator, and scavengers often ate the leftovers. Geologic upheavals of various types added to the disarray of fossils. That is why paleoanthropologists are excited to discover even an individual tooth, jaw, or limb, let alone a complete cranium.

Upon discovering a fossil, a paleoanthropologist asks what, when, and where? *What* defines the species; *when* determines its relative or absolute age; and *where* specifies its location and the climate in which that species lived.

NAMES OF FOSSILS

Each fossil is assigned a name according to a binomial system outlined by Carolus Linnaeus, an eighteenth-century biologist. As an example, consider *Homo sapiens* (*wise man*), the two-part name that refers to modern humans. Those terms identify, respectively, our genus and species, the lowest terms of the taxonomic hierarchy. Going from the most general to the most specific, these are kingdom, phylum, class, order, superfamily, family, subfamily, tribe, genus, and species. A full name specification for a human being would be: *Animalia, Chordata, Mammalia, Primate, Haplorhini, Hominidae, Hominini, Homo, Homo sapiens*.

How is membership in each group decided? The rule is quite simple at the bottom of the hierarchy. A *species* is defined as a group of organisms that can interbreed to produce a fertile offspring.[2] Organisms of different species of the same genus cannot. A mule is a classic example. It is a product of a donkey (*Equus africanus asinus*) and a horse (*Equus ferus caballus*). Even though both species belong to the same genus (*Equus*), their offspring, mules, are infertile. Recent DNA evidence suggests that some *Homo sapiens* interbred with *Homo neanderthalis*, our most recent ancestor.[3] There is also evidence of a viable offspring produced by a Neanderthal mother and a Denisovan father.[4] These are clearly exceptions to the rule that members of different species cannot produce a viable offspring. Biologists have yet to discuss the implications of such interbreeding for defining membership in a given species.

In this chapter, I concentrate on hominins, a tribe of the subfamily of *Hominidae*, most of whose members were discovered during the past few decades. Fundamentally, they differed from chimpanzees in that they were bipedal. As we shall see, other anatomical differences are relevant, some of which were crucial for the evolution of language.

AGE OF FOSSILS

Paleoanthropologists use both relative and absolute measures to define the age of a fossil. Relative measures link a fossil's age to independently established ages of neighboring fossils and/or rocks. Absolute measures rely on physical properties of the fossil and/or neighboring rocks.

A fossil's relative age is determined by its stratigraphic level; that is, the number of strata (layers of earth and rock) above and below its position in the ground. Relative dating makes use of the geological principle that older strata lie below younger strata. Louis and Mary Leakey's research in the Olduvai Gorge in Tanzania provides a good example. They discovered hominins lying in physically distinct strata that were distinguished by the relative amounts of volcanic ash they contained. At the lowest level, they discovered a fossil that came to be known as *Australopithecus boisei*; at the next highest, a more recent fossil of *Homo habilis*; and at the next highest, *Homo erectus*.

Because each stratum was physically distinct, it followed that *Australopithecus boisei* was older than *Homo habilis* and that *Homo habilis* was older than *Homo erectus*. In this instance, stratigraphic dating provided information about each fossil's relative age, but not its actual age. That requires a measure of absolute dating.

The most widely used absolute dating techniques are called *radiometric*, because they use radioactive decay to measure time. An example is carbon dating, which makes use of the fact that many living organisms incorporate two isotopes of carbon: carbon-12, whose atoms have six neutrons; and carbon-14, whose atoms have eight. The atmosphere contains both types of carbon. While an organism is alive, the ratio of those isotopes in its body is the same as it is in the atmosphere. After the organism dies, though, it no longer absorbs carbon-14. The amount that is left in the dead organism decays into carbon-12 at a constant rate. Specifically, that amount is reduced by half every 5,700 years, a value that is defined as the "half-life" of carbon-14. It follows that the time at which an organism died is determinable by the amount of carbon-14 found in its fossil remains.

The same logic has been used to develop other radiometric measures, such as potassium-argon dating, which measures the amount of argon that potassium-rich rocks retain following volcanic eruptions. Whereas carbon dating is suitable for relatively recent events (up to 500,000 years), potassium-argon dating can be used to measure time in the range of 10,000 years to 3 billion years.[5]

Potassium-argon dating was used in the Olduvai Gorge to determine the age of the rocks lying near the fossils that the Leakeys discovered. This measure showed that the age of fossils of *Australopithecus boisei*, *Homo habilis*, and *Homo erectus* were, respectively, 2.3 million, 1.9 million, and 1.6 million years old. Whereas relative dating only provided information about which fossil was the oldest, the next oldest, and the youngest, absolute dating corroborated that order and provided actual dates.

LOCATION OF FOSSILS

Where a fossil is found is easy to determine, but to answer the larger *where* question it is also important to consider the climate in which that ancestor lived and the climate/weather it experienced. That helps us understand the challenges a particular species had to overcome to survive.

Paleoanthropologists agree that hominins evolved in East Africa, but the questions of how and why they evolved don't lend themselves to a simple answer. Just as paleoanthropologists had to modify Darwin's theories about the linear descent of our ancestors, recent information about climate change in that part of the world has influenced traditional theories of the emergence of new species. Instead of responding to a single change of climate in East Africa, our ancestors had to adapt to multiple climate changes in their lifetimes.

CLIMATE CHANGE

One overly simplistic theory, which dominated anthropological thought for most of the twentieth century, is known as the "savannah hypothesis" (or the "East Side Story").[6] It is based on a shift in the tectonic plates under East Africa that caused a dramatic fragmentation of its landscape at the beginning of the Miocene era, about 25 million years ago. The most prominent consequence of that shift is the Great Rift Valley, a huge rupture that runs approximately 4,000 miles and reaches a width of 60 miles. Many lakes appeared in its vicinity, as did the two tallest peaks in Africa: Mount Kilimanjaro and Mount Kenya (both almost 20,000 feet). Figure 3.1 shows the most distinguishing features of the Great Rift Valley.

For millions of years, jungles to the west of the Great Rift Valley benefited from a warm and humid environment. In the east, however, newly formed mountains caused a rain shadow that interrupted monsoons moving in from the Indian Ocean. As a result, the jungles on the eastern side of the Great Rift Valley were replaced by lightly forested savannahs and open grassland. About 5 million years ago, a sharp reduction in air and water temperatures made this area even more arid.

Chimpanzees continued to thrive to the west of the Great Rift Valley, where they obtained food while swinging from one branch to another. On the eastern side, however, our early ancestors had to roam great distances on foot to find food. Supporters of the savannah hypothesis argued that this change of vegetation resulted in bipedalism, a new form of locomotion.

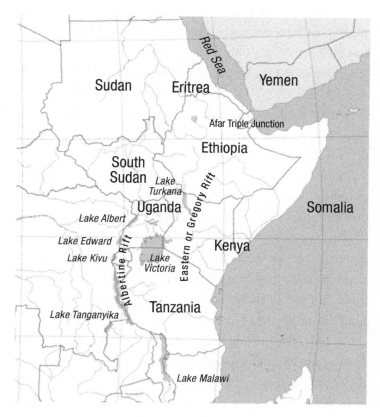

FIGURE 3.1 Great Rift Valley, Eastern Africa. The Great Rift Valley varies in width from twenty to sixty miles and in depth from a few hundred to several thousand meters. The diversity of environments in this region provided an ideal setting for evolutionary change and has been a rich source of fossils.

Davies (2008).

A major problem with the savannah hypothesis is the location of hominin and chimpanzee fossils. Until recently, hominin fossils had only been discovered in Eastern Africa and chimpanzee fossils had not been discovered on that side of the Great Rift Valley.[7] In 1992, however, hominin fossils were discovered in Chad, thousands of miles *west* of the Great Rift Valley.[8] More recently, chimpanzee fossils have, for the first time, been found in eastern Africa.[9]

There is also evidence that many early hominins were not completely bipedal. They spent some of their time walking in the open savannah, but they also climbed trees during the day to avoid predators and slept in trees at night.[10] Another limitation of the savannah hypothesis is its inability to explain changes other than bipedalism. We know, for example, that brain size tripled millions of years *after* our ancestors moved from the jungle to savannah regions.

At best, the savannah hypothesis can explain the early changes in hominins, beginning about 5 million years ago. Recent discoveries by climate scientists have shown that our ancestors had to adjust not only to one new environment—the shift from a jungle to open savannah—but also to variable environments within the savannah. The shift from the jungle to open grasslands was gradual, not abrupt.[11] The changes occurred during a succession of wet-dry periods during the shift to relatively arid environments. As we shall see, two of those periods coincided with the emergence of new groups of ancestors (*Australopithecus* and *Homo*).

To discover those changes, climate scientists have used a variety of formidable tools to analyze many phenomena.

Those phenomena include the type of vegetation our ancestors ate, the composition of the enamel of their teeth, different types of oxygen isotopes in the air, the amount of pollen, and the chemical composition of different levels of silt under the Indian Ocean.[12]

Beginning about 3 million years ago, there were wild fluctuations in the environment near the great Rift Valley. As the climate varied in short pulses every 20,000 years from wet to dry, lakes along the Great Rift Valley filled and drained accordingly. The mountainous landscape made the lakes very sensitive to changes in rainfall. That caused the rapid appearance and disappearance of those lakes at short intervals. The variability of rainfall was especially pronounced 2.6 and 1.8 million years ago, a time at which many ancestors became extinct and new species replaced them.

The picture that emerges from these discoveries is that there were two major shifts in climate in East Africa. The first occurred 3 to 2.5 million years ago. At that time, Lucy (a member of *Australopithecus afarensis*) became extinct and the first members of *Homo* appeared. The volume of the brains of the oldest species of the genera *Homo* increased markedly above that of its ancestors. *Homo habilis* also appears to have been the first species to manufacture simple stone tools.

The second shift occurred 2 to 1.5 million years ago. During that time, *Homo erectus* appeared, with a greatly increased brain volume and a skeleton similar to our own. *Homo erectus*, which introduced a new type of stone tool referred to as Acheulean, was the first ancestor to migrate from Africa. This species thrived in southern Asia, China, and Java.

One paleoanthropologist concluded that, beginning about 5 million years ago, hominins increased their ability to cope with *changing* habitats rather than remaining specialized in a single type of environment.[13] Increases in the variability of climate placed a premium on being nimble and versatile. In addition to being able to cope with a particular environment, it was important to become adaptable to variable environments. That ability allowed our ancestors to succeed not only in the variable climate of the savannah, but also in new areas— eventually, all areas on this planet.

BECOMING BIPEDAL

This review of hominin evolution covers two themes. One is the shift from quadrupedalism to bipedalism, a transition during which our ancestors' brains were barely larger than a chimpanzee's. The other is the enlargement of the brain, a change that occurred millions of years after bipedalism emerged. As noted earlier, each of these shifts coincided with a major change in climate.

To walk upright, an organism's skull has to be centered above its spinal column in an area called the foramen magnum (Latin for "big opening"). What hominins have in common are skulls in which the foramen magnum is located in a relatively central position at the bottom of the skull. Figure 3.2 contrasts the attachment of the spinal cord to the brain in a chimpanzee and a hominin. As shown in figure 3.3, the foramen magnum in a chimpanzee is located toward the

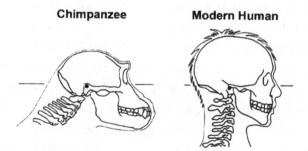

FIGURE 3.2 The connection of the spinal cord and the skull in a chimpanzee (left) and a modern human (right).

Adapted from Jaanusson (2007).

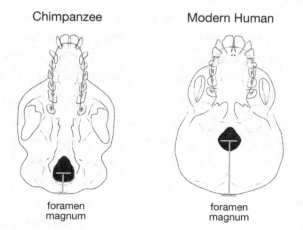

FIGURE 3.3 Base of skulls of a chimpanzee and a modern human.

Adapted from Lewis, Jurmain, & Kilgore (2006), 203.

back of the skull. In hominins, the foramen magnum is more forward.

When our ancestors began to walk bipedally, they had what is referred to as *facultative bipedalism*. They could walk or run, but only for short distances. The ability to walk or run long distances, which is called habitual or *obligate* bipedalism, is rare and took millions of years to evolve.[14] Among living hominins, only humans have this ability, although some extinct members of *Homo* also did. Obligate bipedalism became prevalent at about the same time that the size of our ancestors' brains began to increase.

Other Anatomical Changes. Species that have facultative bipedalism had long arms that helped them swing from one branch to another, and curved fingers and toes that were conducive to climbing. Figure 3.4 shows the skeletons of a chimpanzee, a human, and one of the numerous species that intervened; in this instance, *Australopithecus afarensis*. Aside from its quadrupedal gait, an obvious difference between the chimpanzee and the human is the relative length of the legs and arms. The arms of chimpanzees and *Australopithecus afarensis* are longer than their legs. The reverse is true in humans. Longer legs make it easier to walk.

Other anatomical accommodations that contributed to the transition from facultative to obligate bipedalism include changes in the spinal cord, the shoulders, the attachment of the femur to the pelvis, the pelvis itself, and the feet. Important changes also took place in the spine. In quadrupeds, the spine functions as a flexible suspension bridge to which the body's organs are attached, an arrangement that makes it difficult

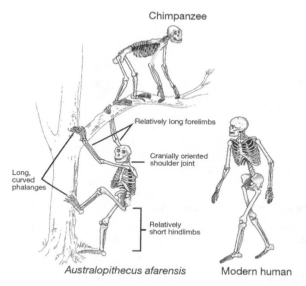

FIGURE 3.4 Drawings of *Australopithecus afarensis*, a chimpanzee, and a human. Although *A. afarensis* could walk bipedally, some of its anatomical features allowed it to climb trees.
Modified from Fleagle (1998).

to walk. To overcome that problem, the center of gravity had to shift downward, from the middle of the torso to the center of the pelvis. As shown in figure 3.5, this meant a change in the shape of the spinal cord, from a "C" shape to an "S" shape. The lower (lumbar) region moved forward and the upper (thoracic) region moved backward. The size and the number of lumbar vertebrae also increased from four to five. An unfortunate consequence of these changes was the transformation of the spine from a load-bearing to a weight-bearing column. That is one reason we suffer from bulging discs and pinched spinal nerves.

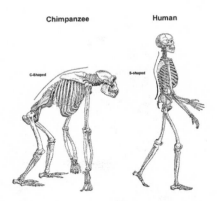

FIGURE 3.5 Spinal columns of a chimpanzee (left) and a human (right).
Adapted from Napier (1967), 63. Reproduced with permission. Copyright 1967
Scientific American, a division of Nature America, Inc. All rights reserved.

Originally, the main function of our ancestors' shoulders was
to enable the climbing of trees and swinging from one limb
to another. As our ancestors began to carry things and throw
objects, their shoulders broadened and they became more adept
at moving their shoulders laterally rather than vertically. Broad
shoulders also increased our ancestors' ability to engage in
long-distance running.[15]

While standing and walking, a biped has to balance the
weight of its torso over both legs. That necessitated a shift in
the manner in which the femur is attached to the pelvis. In
chimpanzees, the femur drops directly from the pelvis to the
knee, whereas in humans and intermediate ancestors (e.g.,
Australopithecines), the femur descends inwardly from the
pelvis to the knee, an angle called the *bicondylar angle*. As shown
in figure 3.6, that angle is minimal in chimpanzees. It is about
17 degrees in humans and Australopithecines.

Chimpanzee *Australopithecus* Human

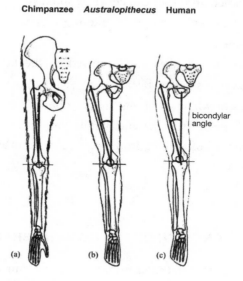

FIGURE 3.6 Angle at which the knee is attached to the pelvis in (*a*) a chimpanzee, (*b*) an *Australopithecus africanus*, and (*c*) a human. Adapted from Shefelbine, Tardieu, & Carter (2002), 765.

In chimpanzees, the pelvis has to support four limbs. Because in bipeds the pelvis only has to support two limbs, its size decreased. A smaller pelvis enhanced a biped's balance while walking and running, but it ultimately caused an obstetric problem. A few million years after our ancestors became bipedal, the size of the brain increased to the point that it became too large to pass through the birth canal. That limited the size of a newborn's brain. Postnatally, the brain grew to its expected size, but its small size at birth had an important consequence: Because the nervous and muscular systems are too underdeveloped when human infants are born, they have to be cradled until they can crawl (at about six months of age). As we shall see in chapter 4, cradling is crucial for language development.

Important changes also occurred in the hominins' feet. Toes became smaller. Because the big toe no longer had to grasp trees while climbing, it lost opposability. Heels and ankles became stronger and arches formed to facilitate the transmission of weight from the heel to the big toe while walking and running.

GROWTH OF THE HOMININ BRAIN

It took approximately 4 million years for our ancestors to fully shift from a quadrupedal to a bipedal gait. During that time, brain size barely exceeded that of a chimpanzee. During the past 2 million years, though, brain size almost tripled. The precipitous growth of the brain is shown in figure 3.7.

The sharp inflection in the brain size function shown in figure 3.7 allows us to distinguish two major groups of ancestors. These are shown in table 3.1, a partial list of hominins that evolved after the split from chimpanzees approximately 6 million years ago. The first group, mainly from the genus *Australopithecus*, had brains whose volume rarely exceeded that of an ape. The second group belonged to the genus *Homo*.

At the time the ape language projects began, only twenty of the twenty-seven species shown in table 3.1 had been discovered. The rapid rise in our knowledge of paleoanthropology is

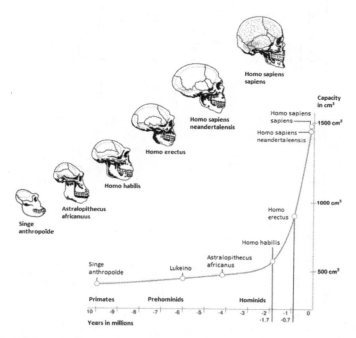

FIGURE 3.7 Increase in brain size during hominin evolution.
van Ginneken, van Meerveld, et al. (2017).

reflected by the rate of discovery of species that were discovered after Project Nim ended (fourteen of the twenty-seven entries).

Table 3.1 also shows the brain size of each species (if known) and when and where that species was discovered. Some of the Latin names of species are easy to understand: *Homo* = man and *Australopithecus* = Southern ape; *Homo habilis* = handy man; *Homo erectus* = upright man. In other instances, a species' name often refers to the location in which it was discovered.

TABLE 3.1 Some hominins that evolved after chimpanzees

Species	Volume (cc)	Lived (mya')	Discovered	Location	Translation	Described by
Sahalanthropus tchadensis	320–380	6–7	2001	Chad	Hope of Life	Beauvilain, as cited in Gibbons (2007)
Orrorin tugenensis	350	5.6–6.2	2000	Kenya	Original Man	Senut et al. (2001)
Ardipithecus kadabba	?	5.6	1997	Ethiopia	Basal Family Ancestor	Haile-Selassie (2001)
Ardipithecus ramidus	300–350	4.4	1994	Ethiopia	Ground Floor	White, et al. (1994)
Australopithecus anamensis	420–500	3.9–4.2	1965	Ethiopia	Lake (Turkana)	Leakey, et al. (1995)
Australopithecus afarensis	400–550	3–3.7	1974	Ethiopia	Southern Ape from Afar	Johanson (2009)
Australopithecus bahrelghazali	400–550	3.6	1993	Chad	River of Gazelles	Brunet et al. (1995)
Australopithecus platyops[†]	350	3.2–3.5	1999	Kenya	Flat Face	Leakey, et al. (2001)
Australopithecus africanus	420–500	2.4–3	1924	E. & S. Africa	Southern Ape of Africa	Dart (1925)
Australopithecus garhi	450	2.5	1998	Ethiopia	Surprise	Asfaw et al. (1999)
Australopithecus boisei[‡]	550	1.1–2.3	1959	Tanzania	Patron of this Fossil Search	Leakey (1959)
Australopithecus robustus[‡]	410–530	1.6–2.1	1938	S. Africa	Heavy-Chewing Complex	Broom (1938)
Australopithecus aethiopicus[‡]	410	2.5–2.7	1985	Ethiopia	Black Skull	Walker et al. (1986)

Australopithecus sediba	420–450	1.98	2008	S. Africa	Karabo	Berger et al. (2010)
Homo rudolfensis	525–700	1.9	1972	Kenya	Lake Rudolf	Alexeev (1986)
Homo habilis	550–687	1.4–2.4	1964	Tanzania	Handy Man	Leakey, et al. (1964)
Homo erectus	780–1225	0.14–1.9	1891	E. Africa & Asia	Upright Man	Antón (2003)
Homo ergaster	700–850	1.4–1.9	1949	Kenya	Working Man	Wood & Collard (2001)
Homo naledi	450–610	0.24–0.34	2015	S. Africa	Star	Berger et al. (2015)
Homo heidelbergensis	1250	0.2–0.6	1908	Germany & Spain	The Muddle in the Middle	Mounier, et al. (2009)
Homo antecessor	1000–1150	0.8–1.2	1994	Spain	Explorer	Bermúdez de Castro et al. (1997)
Homo cepranensis	?	0.8–0.9	1994	Italy	Ceprano, Italy	Manzi et al. (2001)
Homo rhodesiensis	1230	0.12–0.3	1921	Rhodesia	Broken Hill Man	Wood (2011)
Homo neanderthalensis	1300–1600	0.4–004	1829	Germany	Name of Valley	King (1864)
Homo flores	400–475	0.1–0.05	2003	Indonesia	Hobbit	Brown et al. (2004)
Homo sapiens (**archaic**)	1200–1400	0.5–0.2	—	Africa & Europe	Wise Man	—
Homo sapiens (**modern**)	1200–1400	0.2–present	—	—	Wise Man	—

*Millions of years ago.

†Considered by some paleoanthropologists to be a separate genus (*Kenyanthropus*) because of the flatness of this species' face.

‡Considered by some paleoanthropologists to be a separate genus (*Paranthropus*) because of the massive ("robust") size of the jaws of each species.

SPLITTING VS. LUMPING

The large number of species in table 3.1 reflects paleoanthropology's status as a young science. The reader should keep in mind that the names of some of those species are not as fixed as names used in other sciences. Consider, for example, elements of the periodic table. Chemistry now has clear criteria for naming each element. Before those criteria were available, however, alchemists often used different names to refer to the same substance. A piece of metal might sometimes be referred to as gold; at other times, as brass or bronze. As we shall see, paleoanthropologists often differ about what to call newly discovered fossils.

With living species, biologists rely on well-documented norms for a population (for example, height, weight, brain size, and so on). By contrast, the norms used by paleoanthropologists are necessarily more variable, given the fragmentary nature of the relevant fossils. That is why it is difficult to assign fossils to a particular species. For example, opinion is currently divided as to whether a particular fossil of *Homo* should be assigned to *Homo erectus* or *Homo ergaster*. In other cases, there are too few exemplars of a particular species to make an unambiguous assignment (for example, *Sahalanthropus tchadensis* and *Homo rudolfensis*).

Paleoanthropologists have to decide which anatomical features are significant in assigning a particular fossil to a particular species. Should the determination be based on the position of the foramen magnum beneath the skull, its height, its degree of sexual dimorphism, the relative length of an arm or a leg, the thickness of a tooth, the shape of the skull, or the alignment of a jaw? Similar questions arise about behavior: for example, particular forms of bipedalism, the rate of postnatal growth, and tool use.

The tendency to use small differences to define and name a new species is known as *splitting*. "Splitters" often envision new fossils as a new species. "Lumpers" envision a new species as more of the same.[16] As more fossils are discovered, it will become easier to determine whether a particular anatomical feature lies within the normal variation of a species' population, in which case the individual will be lumped with other examples of that species; or whether it is different enough to justify naming it a new species. Stable norms for defining each species will eventually be reached. Barring the discovery of new species, their number will undoubtedly decrease.

For the sake of simplicity, I have organized the entries in table 3.1 from a lumper's point of view. All of the species in the first group, most of which are *Australopithecina*, have a foramen magnum that is centrally located at the base of the skull, implying bipedalism. Another common feature is the relatively small size of each species' brain, barely larger than a chimpanzee's. That is why they are called bipedal apes. Species of the second group, referred to as *Homo*, can be differentiated from *Australopithecina* by three features: they all have upright posture, a relatively large brain, and the ability to manufacture and use stone tools.

Paleoanthropologists have made admirable progress in determining the genealogy of the species shown in table 3.1. Nevertheless, there are gaps. It has been argued, for example, that *Australopithecus anamensis* descended from *Ardipithecus ramidus*, but some think that the interval between the extinction of the latter and the appearance of the former is too short an interval for a new species to evolve (about 200,000 years). A similar question arises in the case of *Australopithecus garhi* and *Australopithecus africanus*. There is still controversy as to which one is our ancestor.

Returning to the debate between splitters and lumpers, some paleoanthropologists argue that labeling the first group in table 3.1 as *Australopithecina* doesn't do justice to some of the species listed there (for example, *Ardipithecus kadabba* and *Ardipithecus ramidus*). It has also been argued that these should be distinguished from various types of *Australopithecus* because their teeth were smaller than those of more recently evolved hominins and because they had opposable toes. Another example is *Kenyanthropus platyops*, which I classified as *Australopithecus platytops*. As their name implies, this species had a flat face, a feature of skulls that did not become common until much later.

Many paleoanthropologists also distinguish between gracile and robust Australopithecines. The former includes *Australopithecus africanus*, *Australopithecus bahrelghazali*, *Australopithecus anamensis*, *Australopithecus sediba*, *Australopithecus afarensis*, and *Australopithecus garhi*. The latter include *Australopithecus aethiopicus*, *Australopithecus boisei*, and *Australopithecus robustus*. Gracile and robust Australopithecines were distinguished mainly by the size of their skulls and teeth, not by the size of their bodies. The strong chewing muscles of robust *Australopithecus* presumably evolved to accommodate their diet of hard-to-process, fibrous, gritty vegetation. Robust *Australopithecus* had bony crests on their foreheads to anchor their strong chewing muscles.

Both types of Australopithecines appeared about 4 million years ago, but gracile *Australopithecus* became extinct about 2 million years ago. Some paleoanthropologists believe that gracile Australopithecines were replaced by *Homo habilis*, from which *Homo sapiens* eventually evolved. There are, however, other candidates. For example, *Australopithecus sediba* has a surprisingly modern hand, whose precision grip suggests that this species might have used stone tools, as did an older ancestor, *Australopithecus africanus*.

Kenyanthropus rudolfensis also deserves our attention. Its cranial capacity was approximately 750 cc. Although none of its remains were associated with stone tools, its large brain suggests that it may have used or manufactured them. With respect to brain size, the brain of *Kenyanthropus rudolfensis* was larger than that of *Homo habilis*. Some paleoanthropologists have used that fact to suggest that *Kenyanthropus rudolfensis* gave rise to *Homo erectus* and, ultimately, *Homo sapiens*. Similar arguments have been made in the case of *Australopithecus garhi*. Evidence of cut marks on the bones of nearby antelope suggest that they were butchered by *Australopithecus garhi*.[17]

Turning to *Homo*, the second group in table 3.1, similar questions arise about the status of *Homo rudolfensis, Homo erectus*, and *Homo ergaster*. There is evidence that *Homo erectus* coexisted with several other early hominin species, including *Homo rudolfensis* and *Homo habilis*. It has nevertheless been argued that *Homo erectus* descended from *Homo habilis*. One suggestion for resolving this puzzle is to postulate some yet-to-be-discovered species of *Homo* that is the ancestor of all of them.[18]

Some paleoanthropologists have argued that *Homo erectus* and *Homo ergaster* are separate species.[19] That may have made sense historically, because *Homo erectus* was discovered in Indonesia in 1894 and it wasn't until the 1960s that the first specimens of *Homo erectus* were discovered in Africa. Eugène Dubois who, in Indonesia, organized the first expedition that looked for human ancestors, discovered what appeared to be the original "missing link."[20] The fossil that was discovered was originally estimated to be only 500,000 years old, but in 1927 its age was placed at about 1.6 million years.[21] Similar fossils were discovered in China. During the past forty years, a growing number of *Homo erectus* specimens have been discovered in Africa that are older than the fossils discovered in China and Indonesia. Figure 3.8 shows the major locations of these fossils.

Was *Homo erectus* a variety of *Homo ergaster* that emigrated from Africa to the Far East, or was *Homo ergaster* a variety of *Homo erectus* that did the same? Given the chronology of when and where specimens of these species were discovered, it's not surprising that paleoanthropologists have yet to agree on an answer. One influential review concluded that the *Homo ergaster* question is "intractable" and that it "remains famously unresolved."[22]

These and related questions about our ancestry will not be settled until more fossils of each species are discovered and norms for each species are established. It is, however, reasonable to assume that one of them was our ancestor. I opt for *Homo erectus* because there is more evidence of tool use by

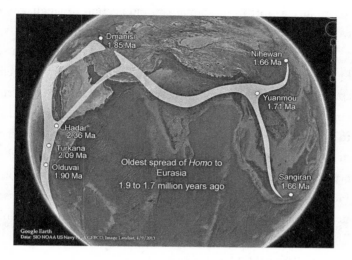

FIGURE 3.8 Key sites and first appearances of *Homo erectus* in dispersal from Africa.

Antón, Potts, & Aiello (2014). Reprinted with permission from AAAS.

Homo erectus than by *Homo ergaster*, and because *Homo erectus* survived longer than any other species of *Homo*.

■ ■ ■

THE INVENTION OF WORDS

A major goal of paleoanthropology is to identify missing links of human descent. My goal in this chapter, which is more modest, is to sift through our recent ancestors to determine which one had the brainpower and the necessary environmental pressures to produce a shift from animal communication to language—words in particular. *Homo erectus* is a plausible candidate.

Homo erectus survived for approximately 1.5 million years, longer than any other species of *Homo*.[23] They were also the first species to emigrate from Africa. Within 200,000 years of their appearance, *Homo erectus* traveled north from East Africa, first to the Middle East and then, traveling along the southern edge of Asia, to Indonesia. Along the way they also settled in southern Russia and China. Such travels pose an intriguing question: How could *Homo erectus* organize its travel groups and devise new destinations within the limits of animal communication? The best answer may come from *Homo erectus*'s solution to a novel nutritional problem it faced.

Homo erectus approached *Homo sapiens* in stature, and its long legs allowed it to walk large distances. *Homo erectus* was also proficient in using stone tools. Crucially, its brain was significantly larger than that of *Homo habilis*.[24] To satisfy the caloric needs of its larger brain, *Homo erectus* required meat as its primary source of food: in particular, meat from large animals such as elephants, wildebeests, giraffes, and so on.

How could *Homo erectus* obtain meat? It lacked weapons to kill large animals, but it could use stone tools to butcher animals that had been killed by other predators or that had died a natural death. Given the need to forage in open grasslands, *Homo erectus* had to adopt a divide-and-conquer strategy. Scouts had to look for dead animals. Having found one, it was necessary to recruit a large group to return to that site to help in butchering it and to confront and scare off other animals that might pick at its remains. In this manner, "confrontational scavenging" became necessary for *Homo erectus*'s survival. Blumenschine, however, noted that scavenging by *Homo erectus* need not have been confrontational: Even passive scavenging in which there were no animals to defend against would have required words. A scout that discovered an animal would still have had to recruit helpers to butcher it.[25]

Two remarkable changes had to occur before *Homo erectus* could take that step. Derek Bickerton, a linguist, proposed that confrontational scavenging required a shift from standard forms of animal communication to words. More recent evidence of scavenging has been reported by Pobiner.[26] Independently, Sarah Hrdy, an anthropologist, observed that such communication required an unprecedented degree of cooperation and trust that would allow the sharing of mental events—a change that allowed *Homo erectus* to become an "emotionally modern human."[27]

According to Hrdy, cooperation in *Homo erectus* was a consequence of intersubjectivity that was instilled by cooperative breeding. *Homo erectus* was the first hominin to engage in that practice. Compared to apes, whose mothers never allow others to care for their young, infants in species that engage in cooperative breeding are cared for and provisioned not only by their mothers but also by other members of their group (*alloparents*). Those infants had to share affect not only with their mothers, but also with other caretakers. Infants who succeeded obtained more attention from their caretakers than those who didn't. That benefit increased the likelihood that they would survive and that, as adults, they would tend to cooperate with and trust their fellows.

There are two parts to Bickerton's theory about scavenging. The factual part requires evidence that it took place. The conjectural part is circumstantial evidence of a new kind of communication that scavenging required. I consider the factual evidence first.

In recent years, paleoanthropologists have used electron microscopes to distinguish two types of marks on fossil bones of large animals: bite marks created by predators that killed and/or defleshed these animals, and cut marks made by hominin tools that sliced through their thick hides. Examples of cut and bite marks are shown in figures 3.9A and 3.9B, respectively. Before 2 million years ago, cut marks lie above bite marks, indicating that hominins accessed those bones only after other animals had scavenged them. After 2 million years ago, bite marks lie above cut marks, indicating that hominins had first access to the bones.[28] Because the animals that *Homo erectus* butchered were too large to have been hunted, the conclusion that *Homo erectus* scavenged large animals is inescapable.[29]

A. Cut Mark **B. Bite Mark**

 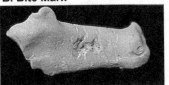

FIGURE 3.9 (*a*) Cut mark from stone tool on antelope leg bone; (*b*) bite mark from lion on a gazelle ankle bone.

Photos: Briana Pobiner.

So much for factual evidence. The conjectural part of Bickerton's hypothesis stems from the need of *Homo erectus* to communicate about dead animals that other members of its group could not see. Bickerton noted the extraordinary degree of cooperation such communication would require, but he did not address its origin. Hrdy's theory of cooperative breeding did. She argued that collective breeding produced the necessary increment in cooperation: one that allowed *Homo erectus* to refer, through sounds and gestures, to objects that others could see.

For Bickerton, the key to understanding scavenging was communication about objects that others *couldn't* see. That was a problem for a scout who had seen a dead animal and who had to communicate its nature and location to members of his group who were too far away to see it. Because of the large distances involved, the scout couldn't rely on pointing to inform his audience of what he had seen and where it was. Instead, the scout had to invent a new form of communication to describe the nature of the carcass and its location. That required what linguists refer to as "displaced reference."[30]

We don't know the form that communication took, whether it was a gesture, sound, or some form of charade. We also don't know if the communication involved sounds that imitated the nature of the carcass to be scavenged, its location, or the nature of rival scavengers. Whatever their form or content, Bickerton argued that the communication used arbitrary words. Words therefore referred to mental entities, representations of absent objects.

The transformation from hominin calls and cries to articulate speech must have taken many years (perhaps hundreds of thousands), but words of some form eventually developed. They could have been used individually or in combinations. Initially, their context sufficed to define their meaning, but problems inevitably occurred as the *Homo erectus* vocabulary increased, as did the length of their utterances. To relieve the burden on working memory, rules for combining words were needed to clarify the meaning of longer utterances (e.g., adjectives before nouns, subject-verb-object order, etc.).

That issue gets us to the topic of grammar, which I discuss in chapter 5. For the moment, however, we should recognize that Bickerton's hypothesis has the merit of explaining the origin of words by the process of natural selection—a significant breakthrough for a theory of the evolution of language.

Mystery persists about how language evolved after words were invented. Was grammar the result of a mutation that occurred about 80,000 years ago, as recently hypothesized by Chomsky,[31] or did it evolve in stages, as others have suggested?[32] Because the only theories about the evolution of grammar are based on cultural variables,[33] a topic that is beyond the scope of this book, I end my discussion of recent ancestors with *Homo erectus*.

As is true of all hypotheses about the behavior of species that we know only as fossils, there is no way of proving that *Homo erectus* uttered the first words. Nevertheless, Bickerton's hypothesis is a heuristic that is more than a "just-so" story because it has the virtue of being about a species that needed words to help it survive. As we shall see in chapter 5, others have suggested that words evolved to facilitate cooperative hunting, to promote pair bonding, to expedite toolmaking, to enhance pedagogy, and so on,[34] but no evidence has been presented that those behaviors were needed for survival. Future research may, of course, question the role that words played in scavenging by *Homo erectus* but currently it appears to be the only viable hypothesis about how natural selection favored the occurrence of words.

■ ■ ■

Summary: Because chimpanzees, our closest living ancestors, are unable to learn language, we must turn to more recent ancestors for clues about its origin. Research by paleoanthropologists who study fossils of our more immediate ancestors provides important clues. Those ancestors, referred to as hominins, include modern humans, extinct species of the genera *Homo* (such as *Homo habilis*, *Homo erectus*, and *Homo neanderthalis*), and older genera such as *Ardipithecus* and *Australopithecus*.

By using tools from physical anthropology, comparative anatomy, genetics, and climate science, paleoanthropologists confirmed Darwin's theory that human ancestors first emerged in Africa and that climate change was a major factor in their evolution. But their discovery of more than two dozen hominins that evolved since we split from chimpanzees, many of which coexisted, disproved Darwin's theory that humans descended from chimpanzees in a linear manner.

Paleoanthropologists identified two major changes in the anatomy of hominins. One concerned their posture; the other, brain size. Posture began to change about 6 to 7 million years ago, during the shift from quadrupedalism to bipedalism. The first major change was the manner in which the spine attached to the skull: in hominins, it is more anterior than in chimpanzees. As compared to chimpanzees, various characteristics of hominins' limbs also changed. Their legs became longer, their shoulders broader, and their heels and ankles stronger. Because it only had to support two limbs, the pelvis (and the birth canal) became smaller.

The volume of hominin brains began to increase about 3 million years ago. Once its size exceeded 1,000 cc (in *Homo erectus*), the relatively small size of the birth canal limited the size of an infant's brain. For humans, this meant that most brain development occurred postnatally. In addition to brain volume, the development of the motor system was limited as well. Human infants, unlike other newborn primates, have to be cradled for the first six months of their lives.

Hominins evolved during a period of major climate change in East Africa. About 7 million years ago, that region was transformed from a relatively flat, forested region to a varied and heterogeneous environment, with mountains more than two miles high and vegetation ranging from desert to savannah. Those appeared along the Great Rift Valley, a giant fissure that is approximately 4,000 miles long.

The appearance of *Ardipithecus*, one of the first bipedal apes, coincided with a major cooling of the environment. According to the savannah hypothesis, bipedalism and the enlargement of the brain resulted from the disappearance of jungles on the east side of the valley. Hominins became bipedal because they had to walk long distances to find food. The savannah hypothesis proved to be an oversimplification. It cannot account for changes during one of the most important periods of hominin evolution, about 1.8 to 2.4 million years ago, during which *Homo* appeared. *Homo habilis* began a tradition of manufacturing stone tools (so-called Oldowan tools). *Homo erectus*, which appeared about 1.8 million years ago, had a brain about three times the size of a chimpanzee's. *Homo erectus* expanded tool technology by adding Acheulean tools. Shortly after they appeared, they also began the first major migration from Africa, to southern Russia, China, and Java.

The size of *Homo erectus*'s brain required a significant increase in caloric intake. Meat was the best way to satisfy that need. Because *Homo erectus* lacked the weapons to kill large animals, they had to find dead animals to scavenge; Bickerton argued that the calories needed to nourish *Homo erectus*'s larger brain made it necessary for that species to scavenge for meat. In that process, a scout had to communicate the discovery of a dead animal. Hrdy, an anthropologist, argued that the practice of collective breeding by *Homo erectus* produced an "emotionally modern" hominin that favored cooperation over competition.

The ability of *Homo erectus* to cooperate allowed scouts who had located a dead animal to solicit help from colleagues who were too far away to see it. Bickerton hypothesized that the need to communicate the nature of the dead animal and its location was responsible for the first words. That hypothesis has the virtue of explaining the origin of words by natural selection, because words were needed for the survival of *Homo erectus*.

Chapter Four

BEFORE AN INFANT LEARNS TO SPEAK

SOMETHING REMARKABLE happens to every infant during her first year that distinguishes her history from that of every other primate. She experiences two *non*verbal relations with her mother that pave the way to language. As one psychologist remarked,

> Those psychologists who believe that humankind became unique by acquiring language are not altogether wrong. But they are not altogether right, either. Before language, there was something else more basic, in a way more primitive, and with unequalled power in its formative potential that propelled us *into* language. Something that could evolve in tiny steps, but suddenly gave rise to the thinking processes that revolutionized mental life . . . . That something else was *social engagement with each other*. The links that can join one person's mind with the mind of someone else—especially, to begin with, emotional links—are the very links that draw us into thought.[1]

This chapter describes the social engagement between an infant and her mother and other caretakers that has been shown to be a foundation of language. That stands in strong contrast to Chomsky's biological approach, which describes the development of language as an asocial maturational process.

Chomsky claimed that language is innate, a manifestation of a biological organ,[2] and that its acquisition is guided by a language acquisition device (LAD).[3] To bolster that claim, Chomsky marshaled an impressive array of evidence. Children go through similar stages while mastering any of the more than 6,000 languages that humans speak. Before they speak full sentences, they can use nouns and verbs appropriately; make negative statements; combine subjects, verbs, and objects correctly; and ask questions—all this and more, without specific instructions. Conversely, children often persist in making mistakes, even when they are explicitly corrected. For example, they pluralize words that are already plural, such as saying *geeses* instead of *geese*, *mices* instead of *mice*, and *deers* instead of *deer*.

Chomsky referred to a child's insensitivity to her linguistic environment as evidence of the "poverty of the stimulus." However, an LAD would be useful only for children who already know some words. The LAD says nothing about how children actually learn those words. If, as Chomsky has argued, language is simply a matter of biology, we might expect an infant to produce words by maturation, just as we would expect her to walk without any external guidance.[4] Indeed, walking and talking are similar in that both activities seem to develop without explicit instruction. However, a child raised in silent isolation would learn to walk, but there's no reason to expect that she would learn to talk. As far as we know, the only way children learn to talk is by conversing with others.

Ironically, Chomsky noticed this more than thirty years ago. Throughout his career he avoided mention of any social factors that contribute to the acquisition of language. But, in what was perhaps an unguarded moment, Chomsky remarked that infants need "triggering events to learn a language," in particular

> a stimulating loving environment in which their natural capacities will flourish. A child that is raised in an orphanage . . . may be very restricted in his abilities. In fact, *it may not learn language properly* [italics added].[5]

The obvious place to study that "stimulating loving environment" is the relation between an infant and her mother.[6] Like other nonhuman primates, human infants form a strong attachment to their mothers that begins at birth. That is evident from their tendency to cling, cry, and seek comfort during stressful situations. Infants are also primed to share and coordinate other emotional and cognitive experiences with their mothers. Such sharing and coordination are needed for the development of language.

During the first few months, human infants and their mothers develop a reciprocal communicative bond, in which they take turns sharing gaze and emotion. That stage is called *intersubjectivity*. Toward the end of her first year, an infant learns to share her mother's attention to external objects. That stage is called *joint attention*. Both stages are uniquely human.

An infant's activities during these stages are represented in what psychologists refer to as "procedural" memory. Unlike "declarative" memory, which is conscious, and which a child uses to recall information once she learns language, procedural learning is not conscious. It is often established by repetitive activities, such as smiling at a caretaker and singing songs. At about six months, an infant begins to play peek-a-boo, an absorbing game that doesn't require any words but which does require the infant to take turns.

It's easy to remember when an infant utters her first words. Episodes of intersubjectivity and joint attention are less memorable because they are not verbal, but that should not detract from the importance of intersubjectivity and joint attention. Language would never develop without those experiences.

Intersubjectivity. Among primates, only humans cradle their infants. As discussed in chapter 3, newborn human infants are the least developed of all primates. At birth, the volume of the infant brain is only about 25 percent of its adult size. In chimpanzees, it is 45 percent.[7] Because the infant's skeletal system is also poorly developed, she can't even crawl until she is about six months old.

An important benefit of cradling is the proximity of the infant's and mother's eyes. That allows them to share each other's gaze, one of many quirks of evolution that contributed to the development of language. Cradling provides ample opportunities for human infants and their mothers to observe and anticipate each other's behavior and to develop patterns of social coordination.

Cradling and, in particular, shared eye gaze contribute to a dramatic example of intersubjectivity right after birth. In a classic experiment, Meltzoff showed that infants can imitate another's facial expression a mere 42 minutes after birth.[8] In that study, an actor protruded his tongue while the infant was sucking on a pacifier. A few seconds later, the actor removed the pacifier. During the next two minutes, the infant protruded her tongue, gradually approximating the actor's movement. Some psychologists objected that the infant's response was reflexive, but that interpretation was ruled out because the baby was sucking on a pacifier when the actor first protruded his tongue and because imitation improved gradually.[9] Photographs of two- to three-month-old infants imitating an actor are shown in figure 4.1.[10]

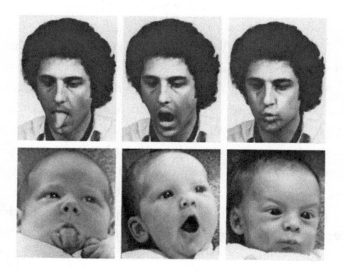

FIGURE 4.1 Responses of two- to three-month-old infants to an actor. Photograph from Meltzoff & Moore (1977), reproduced by permission of A. Meltzoff.

The imitative ability of an infant is all the more remarkable because she has never seen her own face. To imitate, she has to match information from two modalities: her perception of the actor's face and proprioceptive feedback from the muscular activity of her own face. As Meltzoff commented,

> The self can be felt, but cannot be seen. The other's face can be seen but not felt. Yet self and other connect. The other can be understood as like me, at least in the sense that we can do the same acts.[11]

Imitation by infants has been amply confirmed by other researchers. Some studies have introduced delays, lasting days, between the actor's movement and the infant's opportunity to respond. Others have made the task harder by having the actor protrude his tongue at a right angle.[12]

A human infant does more than just imitate. If that's all she did, she would be unable to share the *rhythm* of her mother's attention, vocalization, and affect, and her behavior would be limited to specific snippets of activity she copied. What's missing is a dynamic picture of how the infant *and* her mother interact: specifically, how they elaborate on and coordinate each other's expressions and gestures on a moment-to-moment basis.

To capture those interactions, developmental psychologists have focused on the *dyad* of infant and mother, and how they learn to coordinate each other's behavior. As the infant develops, her mother monitors the infant's emotional state to look for activities they might share. While engaged in those activities, the infant develops *expectancies* about how her behavior affects her mother and how her mother's behavior affects the infant. The fulfillment of those expectancies adds to the infant's ability to discriminate her affect from that of her mother.

Much of the literature on intersubjectivity is based on studies of three- to four-month-old infants. By that age infants are awake long enough to engage in sustained play sessions. To the untutored eye, an infant's repertoire may seem limited. But even though infants don't babble until they are six months old, they can make many sounds (for example, whimpers, coos, and grunts). They may be unable to crawl or stand, but they can smile, move their hands, turn their heads, and so on.

Those activities occur very rapidly. Facial expressions, for example, often change more than once a second, making them hard to document. Recent advances in video technology have overcome that problem. One study compared how a two-month-old infant reacted to immediate and delayed videotapes of her mother's behavior.[13] After mothers and infants were seated in two rooms, they were asked to interact while watching and hearing each other on a video screen. In one condition, the infant and her mother interacted in real time. In the other, the mother's response was delayed by one second. The real-time and delayed conditions alternated seamlessly.

During the real-time condition, the infant often looked at her mother's face, smiled, and expressed positive affect. Those reactions, which alternated with her mother's, were interpreted as preverbal conversation. During the delay condition, however, the infant was puzzled and confused, often looking at the mother with a hesitant and expressionless face. That result was attributed to the infant's loss of a contingent connection between her own and her mother's behavior.

Another study eliminated any feedback from the mother after she played with the infant for two minutes. At that point she was asked to maintain a "still face" for two minutes: that is, to look straight ahead and not react to the infant's behavior.[14] The infant's reaction to the still face was quite predictable. First, she became very agitated. She then tried to restore the reciprocal pattern of activities to which she was accustomed by looking, smiling, and vocalizing. When those attempts failed, she withdrew and turned away from her mother. Her negative affect was often accompanied by gestures that made it seem as if she was straining to reconnect with her mother. Figures 4.2A and B show how an infant reacts during a still-face experiment.

To provide a more dynamic view of reciprocal displays of mother-infant interactions, Beebe used the method of "microanalysis,"[15] which was pioneered by Trevarthen[16] and Stern.[17] While a mother and her three- to four-month-old infant engaged in normal play, they were videotaped by separate cameras. The videotapes were then time-synched and displayed on a split screen. Well-trained raters then assessed those images for expressions of affect.

Microanalysis reveals interactions between a mother and her infant that are difficult, if not impossible, to see in real time. Consider, for example, the synched frames in figure 4.3, which were selected from a study that measured smiling while a mother and her infant gazed at each other during a three-minute session of unstructured play. In figure 4.3A, the mother smiles at her infant. Less than two seconds later, the infant smiles back (figure 4.3B).

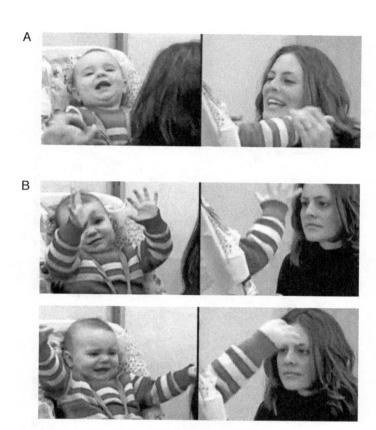

FIGURE 4.2 (*a*) An infant's response during a normal interaction with her mother; (*b*) the infant struggles to recapture her mother's attention during the "still-face" interaction.

Adapted from: E. Tronick. Still Face Experiment. Source: https://www.youtube.com /watch?v=apzXGEbZht0 (Umass Youtube 2007)

FIGURE 4.3 Synchronized frames from videotapes of a mother and her infant. The numbers at the bottom right corner of each frame provide a temporal resolution of behavior to the nearest thirtieth of a second. See text for additional information. Beebe, Cohen, & Lachmann (2016).

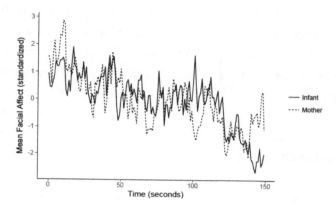

FIGURE 4.4 Mean ratings of mother and infant facial affect during 150-second sessions of 132 mother-infant dyads. Mothers and infants each followed the other's direction of affect changes.

Data obtained from table 1 of Beebe, Messinger, et al. (2016).

Beebe used microanalysis to measure shared affect between mothers and their infants.[18] As shown in figure 4.4, the degree of facial affect they expressed was highly correlated. Mothers' and infants' levels of positive and negative facial expression were high at the beginning of the session, but they decreased gradually as the session progressed, as is normal during play sessions. The mother's and infant's levels of positive, neutral, and negative affect were nevertheless similar throughout the session.[19]

Bateson subjected vocalizations to a similar analysis.[20] She recorded the relation between the utterances of two-month-old infants and their mothers, in response to the mother saying, "What you gonna say?," "Huh?," "Oh my!,"

"You gonna be a good boy today?," and so on. The infant often responded by cooing, grunting, whimpering, and making other infant sounds. Bateson noted a strong correlation between the onsets of the mother's comments and the infant's responses. Because there was little temporal overlap between those utterances, Bateson referred to them as "proto-conversations." That interpretation seems justified because infant and mother alternated their utterances, just as adults do in real conversations.

Beebe and her colleagues extended this approach by recording the onset and offset of vocalizations of three- to four-month-old infants and their mothers, in units as short as 250 milliseconds (msec).[21] Turn-taking was measured by comparing the duration of the infants' and mothers' switching pauses: that is, the time that elapsed between the offset of one speaker's vocalization and the onset of the other's. The duration of switching pauses ranged from 0.1 to 1 second. As shown in figure 4.5, the average duration of pauses was highly correlated.[22]

Each partner paused for a similar duration before taking a turn, in effect adjusting the duration of the pause to approximately match the other's. Those exchanges can be likened to a dance between partners of differing ability who are trying to keep in step by maintaining a mutual rhythmic pattern. Although neither partner dances precisely to a particular beat, they can each vary the duration of their steps. That helps them to synchronize their dancing.

Most people regard the interaction of a mother and her infant as special. Microanalysis shows why. Figures 4.4 and 4.5

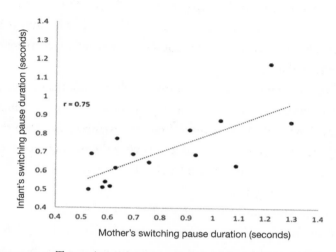

FIGURE 4.5 Turn taking in mother-infant dialogue (infant age: 3.5 to 4 months) as measured by the duration of their switching pauses. Data are based on table 3 of Beebe, Alson, Jaffe, et al. (1988).

illustrate the tight coordination of affect between an infant and her mother. This is something to keep in mind the next time you see a mother sharing love with her infant.

Joint Attention. Only humans know that things have names; that is, that names refer to things. Why humans do, and why other species lack such knowledge, is another way of asking: What are the critical features of language?[23] Grammar is obviously important, but children must understand that words refer to things before they can master grammar.

Joint attention, which is a natural outgrowth of intersubjectivity, helps to answer these questions. Earlier in this chapter, we discussed how a human infant and her mother form an affective bond which surpasses that of any other primate.[24] Intersubjectivity is a necessary step toward language, but it is not sufficient because it does not include any reference to environmental events.

Infants start to crawl at approximately six months of age, at which time they then begin to explore objects, many of which they share with their mothers. In some instances, they hold the objects up to allow their mothers to view them; in others, they point to the objects. The reader should keep in mind that such sharing is nonverbal.

Joint attention allows the infant to learn the names of objects by providing an opportunity for the infant and her mother to verbally share their perceptions of what they see. An example will reveal the intricacies of this deceptively simple type of communication. Suppose that a mother wants to teach her infant how to refer to their cat. While she and the infant are playing with it, the mother says *kitty*. But *kitty* could refer to a multitude of other nearby objects: a chair, a car, a bird, a tree, a dog's tail, and so on.[25] Were it not for joint attention, the infant would have no idea what *kitty* meant or referred to. However, once the "common ground" of joint attention is achieved, it's easy for the child to understand that her mother's attention was focused on the cat.[26]

How is a common ground achieved? Many behaviorists have argued for an associative mechanism,[27] but it has been shown that such mechanisms are hopelessly inadequate.[28] In the absence of distractors, an associative mechanism can predict the occurrence of one stimulus following another, for example, that a light will follow a tone. Prediction becomes difficult, if not impossible, to the extent that visual distractors accompany the light or auditory distractors accompany the tone.

For an associative mechanism to explain word learning, an adult would have to name an object at the precise moment at which the infant looked at it *and* the infant would have to detect the co-occurrence of the adult's utterance and the target of the child's perception. But how would the infant know what that target is? She is just as likely to associate the target with some other feature of the environment the infant notices that accompanies the word.

What's missing from an associative model is a mechanism to insure that an adult and an infant share their perception of an object. That is what joint attention accomplishes.[29,30] First, the infant and the adult look in the direction of an object. They then confirm that they each saw the object. The infant might point at it or smile. The adult might do the same. The second step increases the likelihood that the infant will connect the name the adult provides with the object they jointly perceived. Only then can the infant learn that *kitty* refers to a cat.

This makes joint attention a two-step process: joint viewing followed by social behavior. An example of the second step is shown in figures 4.6. An infant and her mother first share their gaze toward a toy. How do the mother and the infant know that they saw the same object? The child smiles at the mother while the mother points at the toy. That confirms their joint perception.

Gaze Following vs. Joint Attention. Joint attention is more complicated than gaze following, a simpler form of social interaction that has been observed in many other species, such as

FIGURE 4.6 Mother and child jointly attending to a toy and engaging in a social interaction that confirms that they both saw the object.
Photo: Tom Lenk
Source: http://www.airbnetwork.org/pictures/PC130941.ORF.jpg

dogs, wolves, goats, birds, and even tortoises.[31] Gaze following is adaptive because it helps to detect predators and locate food resources. For that reason, it has been described as a socially facilitated orienting response. Humans also engage in gaze following, but joint attention is uniquely human.

To appreciate the difference between gaze following and joint attention, imagine walking with a friend and looking up at the sky to observe a passing airplane. It is likely that your friend would do the same. Or, if you were both at a theater watching a play, you would be sensitive to each other's gaze as you tracked the movements of various actors. But even though you gazed at the same object, you couldn't say that you both saw it. In these examples, gaze following is simply *parallel* looking. By contrast, joint attention means "knowing something together."[32]

That implies a social interaction in which each party acknowledges that he or she perceives the same object (cf. figure 4.6).

Research has shown that apes can infer what another ape sees, but only in competitive situations involving a primary reward. Consider, for example, a dominant and a submissive chimpanzee, each housed in separate rooms that are separated by a common area. It should come as no surprise that the subordinate refrains from grabbing food when an experimenter places some in the common area between the rooms.[33] That's because immediate access to food is a dominant chimpanzee's prerogative.

A simple change of procedure produces the opposite result. The subordinate grabs the food without hesitation if a visual barrier is placed between the dominant chimpanzee and the food. That observation led to the claim that the subordinate inferred what the dominant could see or, in this instance, couldn't see.[34] Others have attributed this outcome to a chimpanzee's memory of earlier competitions for food.[35]

Whether or not a chimpanzee can infer what another sees, there's no evidence that they engage in joint attention, either in the wild or in laboratory settings. In the most ambitious experiment to address this issue, "encultured chimpanzees" and children were encouraged to imitate a trainer who performed simple tasks with various toys.[36] Analyses of videotapes of each species' behavior provided ample evidence of joint attention (and pointing) in children, but not in chimpanzees.

Why the absence of joint attention in a species that can follow gaze and, according to some, infer what another sees? An obvious reason is a chimpanzee's poorly developed ability to cooperate and to share. As far as I'm aware, there is no evidence that a chimpanzee will trade one or more of its possessions for another's, either in their natural environment or with a human caretaker (say, a bunch of grapes for a banana, or vice versa, or one toy for another). By contrast, there's ample evidence that children spontaneously share food or toys with others.[37] With the exception of a mother who occasionally allows her infant to steal scraps of food, there is no evidence of spontaneous sharing in chimpanzees.[38]

After reviewing the social behavior of chimpanzees and young *Homo sapiens*, Tomasello hypothesized that human infants, but not chimpanzees, "share intentionality." That is why chimpanzees can't engage in joint attention.[39] *Shared intentionality* refers "to the ability to participate with others in collaborative activities." As an example, Tomasello cited an activity in which an infant and her mother build a tower of blocks. After one lays down a block, the other follows suit. While working together, the mother and infant often smile and gaze at each other. Rather than working for a concrete reward, each individual's motive is to share affect while building the tower.

I agree with Tomasello's hypothesis that shared intentionality is uniquely human, but I would add that an infant's intersubjective experience, which is also uniquely human, is an essential precursor. By the age of three months, infants engage in bouts of mutual smiling with their mothers. They also take turns while vocalizing, and point to and draw their mothers' attention to particular objects. There is no evidence that chimpanzees do so. It is doubtful that shared intentionality would emerge without the strong affective bond that an infant forms with her mother during intersubjectivity.

More recently, Higgins proposed the concept of "shared reality" as a broader version of Tomasello's hypothesis: "Shared reality is the experience that you have an inner state about something that is shared by others (e.g., a feeling or belief or concern about something)."[40] Shared reality captures more of the uniqueness of human infants than shared intentionality but, as it is presently defined, shared reality begins earlier than Higgins and Tomasello propose (at approximately six months). As we've seen, shared reality begins at birth.

The absence of joint attention in chimpanzees explains why they aren't able to learn language. As I mentioned in chapter 2, they fail not because they can't learn grammar, but because they can't learn the declarative function of words. Chimpanzees are only able to use words as imperatives, to obtain primary rewards.

The manner in which a human infant and Nim learned to sign *kitty* and cat, respectively, illustrates this difference (cf. chapter 2, figure 2.3, figure 2.4A and B). For Nim, signing *cat* was not part of a dialogue. He was uninterested in any verbal acknowledgment from his teacher. By contrast, declaratives are truly conversational. Their only function is to transmit information about a mutually interesting object. There is no evidence that a chimpanzee's utterances are declarative or, more generally, that the utterances of any animal are declarative. That difference in the use of declaratives is significant. If human communication were limited to imperatives, language would never have evolved.

Given our knowledge of intersubjectivity and joint attention, both of which were discovered after Project Nim, it is easy to understand why that and similar projects failed. Chimpanzees lack the affective and cognitive foundations necessary for language. Anyone considering a similar project would do well to consider their importance. It is inconceivable that an animal could learn names without the ability to engage in joint attention.

From Joint Attention to an Infant's First Words. For joint attention to work, the infant must differentiate when an adult turns toward an object and when the adult sees it. That ability appears to develop toward the end of the infant's first year. This was shown in an experiment that measured gaze orientation in nine-, ten-, and eleven-month-old infants who were trained to look at the experimenter at the start of each trial. In half of the trials the experimenter's eyes were open; in the other half they were closed. After the infant looked at the experimenter, the experimenter turned toward a target. How often did the infants turn?

Ten- and eleven-month-old infants were much more likely to look at the target when the experimenter's eyes were open than when they were closed. By contrast, the younger infants were equally likely to follow the orientation of the experimenter's head, whether or not his eyes were open. Thus, as infants approach their first birthday, they begin to sense when an adult can see objects that lie in the direction of his head orientation.[41]

Another experiment showed that one-year old infants were sensitive to an adult's interest in nearby objects.[42] They were likely to point at that object if the adult looked at it *and* made a positive comment about it. They were less likely to look at it if the adult ignored them while looking at the object, or if he commented about it without actually looking at it.

These and other experiments show that one-year-old infants are uniquely primed to benefit from joint attention when they begin to learn names. What remains to be shown is the influence of joint attention on an infant's *initial* vocabulary. As the infant gets older, the importance of joint attention diminishes because she can learn new words while verbally interacting with her parents and with other language users. The critical question is this: How does an infant *begin* to learn words?

It is well established that infants rarely learn to produce words before their first birthday and that (in Western cultures, at least) vocabulary increases at an accelerated rate from that time.[43] It is also established that an infant's growth of vocabulary is facilitated by the degree of joint attention to which she is exposed while learning words.[44] In each instance, there was a positive correlation between the extent to which young infants (eight to fourteen months old) engaged in joint attention with the experimenter and the size of their vocabulary at a later age (eighteen to thirty months). What caused that correlation is difficult to determine because studies typically provide multimodal cues (for example, vocalizing *and* pointing to particular objects).

One experiment that controlled for those variables measured vocabulary growth solely on the basis of an infant's ability to follow another's gaze. The subjects were twelve-, fourteen-, sixteen-, and twenty-four-month-old infants. After making eye contact with the infant the experimenter turned silently to one of two targets and fixated on it at the beginning of a trial. On each trial, the experimenter measured the frequency and duration with which infants looked at the target and the extent to which they pointed to it. After each session, the infants' vocabulary was assessed by a standard inventory.[45]

At the age of two years, the average size of the vocabulary of infants in a control group, who did not participate in this study, was approximately 310 words; of infants who followed the gaze of the experimenter, 480 words; and of infants who followed the gaze of the experimenter and pointed to the target, 520 words. These results clearly delineate the contribution of joint attention and pointing to the growth of vocabulary.

Developmental psychologists have also identified other aspects of joint attention that influence vocabulary size. For example, the growth of vocabulary increases more rapidly in infants whose mothers follow their pointing to new objects than it does for infants whose mothers direct their infant's attention to new objects.[46]

Collectively, these experiments confirm the vital role of joint attention in establishing a child's vocabulary. Even as a thought experiment, however, it is inhumane to think of methods for controlling its influence. Such experiments would be tantamount to eliminating the "stimulating loving environment" that Chomsky proposed as necessary for language to flourish. There are, however, two "natural" conditions which approximate that condition: children raised in orphanages and autistic children.

Orphanages. Toward the end of the twentieth century, the Romanian dictator Nicolae Ceausescu outlawed abortion and contraception and decreed that women should bear at least five children. Ceausescu also proclaimed that the state could provide better care for infants than their parents could. As a result, more than 150,000 children and orphans were abandoned because their parents were too poor to raise them or were absent. In many cases, the care provided by the state was very primitive.

Institutionalization had many adverse physiological and psychological effects, many "breathtakingly awful."[47] For our purposes, the most relevant findings were that (1) the longer an infant was institutionalized, the more severely retarded the development of her language; (2) transfer of an infant from an orphanage to foster care within one year facilitated normal recovery of linguistic function; and (3) in many instances, full recovery of linguistic ability was not obtained if a child was transferred to foster care after more than two years in an orphanage.[48]

Autism. Autism is a behaviorally defined neurodevelopmental disorder that affects as many as one in a hundred infants. Its most obvious manifestation is social, and it can usually be detected by the age of three years. Autistic infants exhibit less eye contact and turn-taking than normal children do. They also spend little time pointing at things[49] and tend to have poor social understanding. Especially significant are large deficits in joint attention.[50] Not surprisingly, language development is retarded in these children. Autistic children rarely begin to produce words until they are two years old.[51]

Albeit for different reasons, the social development of infants who were raised in orphanages or who are autistic is limited.

In orphanages, there is neglect on the part of caretakers. For autistic children, neural factors attenuate their ability to relate emotionally with their caretakers. In both instances, however, a lack of social skills interferes with language development. Both types of children are poignant reminders of the importance of a "stimulating loving environment" for the development of language, which Chomsky briefly recognized but later ignored. I elaborate on the importance of social factors during the linguistic development of normal children in chapter 5.

■ ■ ■

Summary: Two nonverbal relations between an infant and her mother—one emotional, the other cognitive—have been shown to be critical for the development of language. Both occur during an infant's first year, before the infant speaks her first words. The first, a dyadic relation called intersubjectivity, is based on sharing affect. The second, a triadic relation called joint attention, is based on the ability of an infant and her mother to share their perception of an external object.

Intersubjective relations are facilitated by cradling. All infants are cradled until they are able to crawl (at approximately six months). While being cradled, infants have a close view of the mother's face, of her eyes in particular. Shared eye gaze allows mother and infant to take turns sharing affect.

To the naked eye, expressions of affect are often difficult to detect because of their fleeting nature. The method of microanalysis has been used to study the dynamics of intersubjectivity. To perform a microanalysis of visually shared interactions, infants and mothers are videotaped by separate cameras while they are engaged in play. The videotapes are then time-synced and presented on a single monitor that allow raters to assess an infant's and her mother's affect on a second-to-second basis. Developmental psychologists have also studied the coordination of preverbal sounds that infants make (whimpering, cooing, grunting, and so on) in response to the mother's vocalizations.

Visual and auditory expressions of affect have been shown to be highly coordinated in studies of smiling, head turning, moving limbs, touching, vocalization, and so on. Experiments in which that coordination was interrupted reveal a deleterious effect, as, for example, when a mother momentarily exhibits a "still face" while interacting with her infant. During the still-face interval, infants immediately begin to struggle to recapture the reciprocity of affect they normally experience.

Even when cradled, infants point to objects of interest in their environment. When they begin to crawl, they often pick up objects and show them to a parent. By the time they are six months old, they begin to look at objects jointly with a parent. Critically, they engage in some form of social behavior after detecting that a parent looked at that object, often pointing to it or holding it up for the parent to see.

That sequence, which is referred to as joint attention, is critical for infants to learn the names of the objects to which they attended. Once the infant and the parent share a common ground of perception, the infant can readily learn to name the object, in response to hearing a parent utter its name. That joint attention contributes to the size of an infant's vocabulary has been demonstrated in studies showing a correlation between the degree of joint attention and pointing and the size of the infant's vocabulary at eighteen, twenty-four, and thirty months.

Low levels of joint attention, such as are experienced by children raised in orphanages and autistic children, have been shown to have an adverse effect on vocabulary size. In the former instance, joint attention is low because of minimal contact with a caregiver. In the latter instance, neural factors attenuate the infants' ability to relate to social cues. Even though the reasons for social neglect are different in each instance, both unfortunate situations demonstrate the importance of intersubjectivity and joint attention in the development of language.

Chapter Five

THE ORIGIN OF LANGUAGE, WORDS IN PARTICULAR

ALL NORMAL human children can learn any of the more than six thousand different languages people speak on this planet. As we saw in chapter 2, no ape can. With lots of training, apes can learn to use some gestures and symbols, but only imperatively, as a means to obtain rewards. Language would never develop if that were its only function.

In their natural environment, the signals of apes, like those of all animals, are innate, involuntary, unidirectional, and immutable. Their only function is to express emotion, such as joy, anger, fear, seeing food, encountering a mate, and so on. Those signals can influence another's behavior, but not another's thinking.

By contrast, the words that children learn are voluntary and flexible. Beginning with their first utterances, children readily learn to share information, conversationally, about how a thing is named. After a child learns to name something in view, she learns to refer to things that are not. When she does, the listener has to think about what the child named.

The difference between animal and human communication is also reflected in the number of signals they each produce. In animals, that number rarely exceeds twenty and it does not vary with age. Human vocabulary grows with age during the first few decades of a person's life. A two-year-old child knows approximately 300 words; a five-year-old approximately 5,000; a minimally educated adult at least 15,000; and a college graduate at least 50,000 words.

Children not only learn the meanings of many words, but they also learn to express new meanings by combining words. Animals cannot combine any of their signals to vary their expression of emotion. There have been few studies of gestures by apes in the wild. One study of 2,321 chimpanzee gestures[1] reported the following types: acquire object/food, climb on me, climb on you, contact, follow me, initiate grooming, mount me, move closer, reposition, initiate copulation, initiate genito-genital rubbing, travel with me, move away, stop behavior. Although there were reports that the strength of these gestures varied with the distance between members of a pair, there was no evidence that they were ever combined. Language and animal communication therefore differ in two respects. Language allows people to name things with words. It also provides a basis for creating new meanings by combining words.

Chomsky recognized the significance of those differences early in his career and developed theories to accommodate them. His theories differed fundamentally from those of traditional linguists. Chomsky rejected their structural approach, the goal of which was simply to describe specific features of different languages (French, German, Russian, Chinese, and so on).[2] In their place, Chomsky proposed that all languages share a Universal Grammar, a set of innate rules that can be applied to any language for generation of meaningful sentences.

Although the details of Universal Grammar changed throughout Chomsky's career, its purpose remained the same. It provides innate knowledge that children use while learning a language and rules that all languages share. In both instances, Universal Grammar is implemented as a set of innate computational rules that works like the operating system of a computer. Its purpose is to generate a set of well-formed sentences from the input of any language, and to discriminate well-formed and ill-formed sentences.

Universal Grammar transformed the current practice of linguistics, but, despite Chomsky's claim otherwise, it provided little insight into the evolution of language. The goal of this book is to show how our ancestors broke the confines of animal communication by using words. Mention of Chomsky may therefore seem irrelevant, especially because he had little to say about the origin of words. Nevertheless, his relevance is inescapable. Chomsky's theories of Universal Grammar have not only dominated linguistics for more than fifty years, but they have also had a profound influence on how many linguists perceive the functions of language.

According to Chomsky, the primary function of language is thinking. Communication is secondary. I think the opposite. Chomsky argues that thinking begins privately in the form of "computational atoms"[3] ("word-like but not words";

see following discussion). I believe that thinking occurs only after a child learns to name things conversationally.[4] A major source of our differences stems directly from Chomsky's current model of Universal Language.

The complexity of those models has decreased progressively during Chomsky's career. Some examples are phrase structure grammar, transformational generative grammar, government and binding theory, and principles and parameters. In each instance, he identified language as a property of individuals rather than of particular groups.[5] That made language amenable to evolutionary study. His most recent model, the strong minimalist thesis, contains but a single recursive operation, called *merge*,[6] a rule I will explain shortly.

In reducing complexity, Chomsky clearly had evolution in mind. As he put it, "despite claims to the contrary . . . there has always been a strong interest in the evolution of language."[7] The reduced complexity of Universal Grammar allowed Chomsky to hypothesize that the evolution of grammar, and hence language, owes its existence to a single mutation—one that produced a "slight rewiring of the brain"—approximately 80,000 years ago.[8]

That is a remarkable hypothesis. How could the complexities of language and its origin result from a single mutation? Equally remarkable is Chomsky's strict focus on grammar. Words also had to evolve. Chomsky recognized that words are "the minimal meaning-bearing elements of human language. [They] are radically different from anything known in animal communication systems." Yet, he thought that "their origin is entirely obscure, posing a very serious problem for the evolution of . . . language."[9] A "serious problem" is an understatement. As Chomsky frequently reminds us, the basic property of language is the ability to generate an infinite number of meanings from a finite set of words. Everyone agrees with that, but without words, there would be no new meanings to generate. As compared to Chomsky's ingenious models of grammar, words have suffered benign neglect.

In asking about words, I'm not trying to detract from the importance of Chomsky's fundamental contribution to our understanding of language. Were it not for the concept of Universal Grammar, we would still be stuck with structural and behaviorist accounts of language, limitations that Chomsky brilliantly exposed[10] (see chapter 1). The use of words, however, constitutes an equally profound break from animal communication. For the first time, it allowed some of our hominin ancestors to use names to refer to objects. No animal has done that.

Although Chomsky is deservedly famous for his insight about the unlimited number of meanings that different combinations of words can generate, he missed a golden opportunity to highlight another basic feature: Just as there is no limit to the number of sentences we can create, there is no limit to the number of names people can invent to identify features of their environment. Human culture has benefited from both abilities.

Chomsky's neglect of words leaves us with half a loaf. The concept of Universal Grammar, which says nothing about the origin of words, applies to the language of young children and adults. What we don't know is whether Universal Grammar reached that state gradually or abruptly. Chomsky opts for the latter but, as I argue toward the end of this chapter, his justification for that position is weak. That would limit Chomsky's contribution of the nature of Universal Grammar to language as used by young children and adults. That's a monumental achievement, but it doesn't concern the evolution of grammar.

Without grammar, there would, of course, be language. However, it is important to recognize that, both phylogenetically and ontogenetically, learning to use words is also an extraordinary development. We have seen that Chomsky attempted to account for the origin of grammar by appealing to a mutation. Chomsky didn't allow himself that luxury in the case of words. As one linguist commented, attributing words to a mutation would be adding "another serendipitous mutation [which would be] too much to swallow."[11]

This chapter addresses some of the consequences of Chomsky's neglect of words and suggest how words could have been naturally selected. My goal is to reclaim words from the obscurity to which Chomsky relegated them and to show how they allow people to share knowledge and exchange thoughts.

Showing how words could have evolved is the first step in explaining the evolution of language. Chomsky claimed to have done that by attributing grammar to a mutation. Unfortunately, that claim is empty, not just because biologists refer to a mutation of that magnitude as a "hopeful monster"[12] (a polite way of saying it's a miracle), but because it fails to bridge the gap between animal communication and language. Evolution is a continuous process, whether by the inexorably slow process of natural selection or by the quantum leap of a mutation. Either way, for Chomsky, grammar is a mutation that lacks a biological anchor.

My difference with Chomsky, however, has to do with words, not grammar. To some extent, it stems from different types of evidence. Chomsky relies on introspective evidence from adult humans: a top down-approach. My approach is to use bottom-up evidence, from research by paleoanthropologists, who suggested a basis for the origin of words in a recent hominin ancestor (cf. chapter 3), and from developmental psychologists, who have shown how *pre*verbal emotional and cognitive communication between an infant and her caregiver is necessary for the development of language (cf. chapter 4).

To resolve our differences, it will be helpful to return to Darwin's theory that language was naturally selected from animal communication, and to a dissenting view by Alfred Russell Wallace, who, ironically, was the cofounder of the theory of evolution.

WALLACE'S PROBLEM

Like Darwin, Wallace recognized the extraordinary challenge that language posed for the theory of natural selection. Initially, Wallace agreed that language, like all other mental functions, was a product of natural selection.[13] However, he changed his mind twenty years after the publication of *The Origin of Species*. Wallace asked how man's "superior intelligence" could result from natural selection, a process that would expand a creature's powers only to the point at which it has an advantage over the competition in the struggle for existence. Specifically, Wallace wondered why humans have "a large and well-developed brain quite disproportionate to his actual requirement."[14]

The human brain is in fact much larger than is needed for its owner to survive and reproduce. It is also a greedy consumer of costly calories. Wallace could see no problem solved by language that could not be solved without it; that is, no problem for which natural selection might have produced language. After expressing his concern to Darwin, Darwin replied, "I hope that you have not murdered too completely your own and my child."[15]

Darwin never addressed the substance of Wallace's defection. That's why the evolution of language has come to be known as "Wallace's Problem."[16] Had Wallace distinguished words and grammar, he would have recognized that the basic evolutionary problem was not grammar, but how words might have been selected for from animal communication. My reading of Wallace's problem suggests that the best answer would be a mechanism in which natural selection favored words: specifically, one that showed words were selected for the survival of a particular species.

Chomsky is one of the few linguists who recognized Wallace's problem. His answer, the strong minimalist thesis,[17] was a mutation that simplified the "language phenotype" to *merge*. That answer seems implausible for a variety of reasons. Chomsky never specified what features of animal communication mutated to produce the language phenotype. Equally problematic is Chomsky's failure to specify how *merge* contributed to our survival. That failure disqualifies it as a solution to Wallace's problem.

BICKERTON'S THEORY OF THE
ORIGIN OF WORDS

A recent theory by Bickerton provides a more plausible answer.[18] Bickerton was explicitly concerned with the origin of words and with Wallace's problem. His theory is based on the linguistic principle of displacement, a design feature of language that allows us to refer to events that are not physically present.[19] That feature requires the speaker and the listener to represent events: that is, to *think* about events that neither party can see. By breaking the connection between the here and now, displaced reference allows a word to become an independent unit of thought, free to refer to past, future, and even imaginary events.

Bickerton asked if there were specific circumstances in which one of our ancestors had to use displacement to survive. To answer that question, he conjured up the ecology, nutritional needs, and cognitive capacities of *Homo erectus*, a species that evolved almost 2 million years ago. Because of extreme climate change at that time, *Homo erectus* lived in a relatively dry environment of open grassland. Its brain size was more than twice that of a chimpanzee's. To satisfy its caloric needs, meat became a primary source of food for *Homo erectus*.

As explained in chapter 3, *Homo erectus* had tools sufficient to butcher large dead animals, but they lacked weapons to kill them. Instead, they first had to find a dead animal that had either died of natural causes or had been killed by a predator. They then had to recruit helpers to butcher the dead animal and to ward off other scavengers (e.g., hyenas). To feed their large brains, the resulting practice of scavenging was essential for *Homo erectus* survival.

Bickerton's theory has two parts: one factual, the other conjectural. How do we know that *Homo erectus* engaged in scavenging? Paleoanthropologists have shown that fossil bones of fauna of that time displayed two types of clues: bite marks from predators that defleshed them and cut marks from hominin tools that scavengers used to cut through their thick hides. Analyses of the distributions of those marks provided evidence that *Homo erectus* often got to dead animals before other scavengers did.[20] Because many of the butchered animals were too large to have been hunted, the conclusion that *Homo erectus* scavenged is inescapable.

Having located a dead animal, how did a scout recruit distant followers to fend off rival animal scavengers and to help in butchery? Thanks to newly formed cooperative habits of *Homo erectus* (see the following discussion), Bickerton proposed that they were already using sounds and/or gestures to refer to objects that were physically present, and perhaps even to refer to footprints or droppings of absent animals. To recruit followers, though, the scout had to communicate the nature of a carcass and its location, things that were far outside the sensory range of potential followers. That could only be done by displaced reference.

The first words, or, as Bickerton called them, "proto-words," necessarily referred to mental entities, representations of absent objects.[21] The form of that communication is a matter of speculation: perhaps mimetic gestures or sounds imitating the nature of the carcass to be scavenged, its location, or the nature of rival scavengers. Some authors have suggested that proto-language consisted of holophrases: that is, groups of proto-words, such as *That's a dead elephant, let's feast on it.*[22] Whether or not *Homo erectus* communicated with proto-words in isolation or in holophrases, it should not detract from the importance of Bickerton's point that the message was arbitrary and that it used displaced reference. Eventually, the vocal modality came to prevail, leaving hands and eyes free to go about their more important functions.

That transformation probably took hundreds of thousands of years, during which the vocabularies of proto-words grew, some of which undoubtedly occurred in combination. Proto-words would have been strung, in no particular order, like beads on a string, relying on context to make sense. That had two consequences. Proto-words that were used repeatedly in combination with other proto-words acquired lexical (combinatorial) status. Sequences of proto-words also created pressure for syntax.

For scavenging to work, *Homo erectus* had to engage in an unprecedented degree of cooperation. Bickerton did not comment on the source of that cooperation but, as discussed in chapter 4, Hrdy hypothesized that it was a consequence of intersubjectivity instilled by cooperative breeding. She argued that cooperative breeding was a crucial factor in transforming *Homo erectus* into an "emotionally modern" human,[23] a creature who cooperated with his peers to a much greater degree than chimpanzees did with other chimpanzees.

Compared to apes, whose mothers never allow others to care for their young, infants in species that engage in cooperative breeding are cared for and provisioned by other members of their group (*alloparents*) in addition to their mothers. A mother would never engage in cooperative breeding or share care of her offspring unless she trusted members of her group.

A collectively bred infant had to learn to share affect not only with her mother, but with other alloparents as well. Infants who succeeded would obtain more attention from their alloparents than those who did not. That benefit increased the likelihood that they would survive and that, as adults, they would tend to trust their fellows and understand them. There is considerable evidence that *Homo erectus* was the first hominin to engage in that practice.[24]

Taken together, Bickerton's and Hrdy's theories suggest that the convergence of four factors created ideal conditions for the emergence of words in *Homo erectus*: the need for calories, inter-subjectivity, joint attention, and the ability to refer to absent objects. Once displaced reference took root, the communicative worlds of apes and *Homo erectus* qualitatively diverged. A small number of innate, unidirectional, involuntary utterances were replaced by a slowly expanding vocabulary of words that were used conversationally.

Intersubjectivity contributed to the emotional develop-ment of *Homo erectus* infants beyond that of apes because of the strength of an infant's rich emotional and communica-tive bond with her mother. Joint attention added a cognitive component to that mix because it is the first instance in which an infant shares her mental state about external objects with another person.

It is, of course, not possible to confirm that the sequence of preverbal stages of language we observe today occurred in our ancestors. However, given the universal sequence of these stages we currently observe, it is unlikely that they appeared in a different order. It's difficult to imagine the occurrence of joint attention without a foundation of intersubjectivity, or of learning words without a foundation of joint attention.

It is nevertheless important to note that neither intersubjectivity nor joint attention was sufficient for the emergence of language. Even if infants engaged in proto-conversations with their mothers (cf. chapter 4), an early manifestation of intersubjectivity, there is no reason to assume that words about objects outside one's sensory range would follow. The same is true of joint attention, which made it possible for our ancestors to comment about mutually interesting events that were in view (e.g., animals, plants, particular people, and so on).

Such exchanges are useful, for example, when naming visible peers, teaching a novice to make stone tools, warning someone about a snake, and other matters of the here-and-now. Nevertheless, those conversations are trivially narrow when compared to those which refer to objects that are not immediately present. They allow people to talk about past, future, and imaginary events, all without any knowledge of grammar!

Words did not achieve linguistic status until they could refer to physically absent objects. It is one thing to use a word to refer to an object that a speaker and a listener can see, and another to refer to that thing in its absence. It is only in the latter case that the speaker and the listener have to think about the relevant event. That form of communication allows words to exist free of time and space. It also allows individuals to mentally combine thoughts about different events. Were it not for displaced reference, there would be no expression of imagination, no possibility of thinking about other words. That is why displacement is an essential step in the evolution of language.

Bickerton's theory about the origin of words is the most erudite and plausible explanation of which I am aware. As a basis for solving Wallace's problem, it is more appealing than other theories about the origin of words because it is based on the assumption that the first use of words, concerning the availability of a remote source of meat, was needed for survival. Other theories cannot claim a need for survival: for example, those that have argued that language evolved to expedite toolmaking,[25] to promote pair bonding,[26] to facilitate cooperative hunting,[27] to promote mother-child communication,[28] as a substitute for grooming,[29] to facilitate a female's evaluation of a male prior to mating (originally, Darwin's Caruso theory),[30] or to enhance pedagogy.[31]

The problem with such theories is that none of them state why language was needed for a species to survive. In the case of toolmaking, *Homo habilis* began the Oldowan tradition of making stone tools approximately 2.5 million years ago. That tradition continued without language for about a million years. The same is true of Acheulean toolmaking, which began about 1.5 million years ago. Pair bonding flourished for millions of years without language, as did cooperative hunting, mother-child communication, and grooming. Contrary to Darwin's Caruso theory, humans are able to attract mates without singing. Words would undoubtedly have facilitated all of these activities, but the species in question survived without them. Pedagogy, which requires reference to specific features of the environment, could only occur *after* the evolution of language. Imitation and emulation—behavior about the here-and-now—clearly predated language, but pedagogy is required for instruction that assumes a shared knowledge of names.

CHOMSKY'S VIEW OF THE ORIGIN OF WORDS

Chomsky denies that proto-language ever existed. As we've seen, he argued that language emerged full blown about 80,000 years ago.[32] We've also seen that Chomsky does not have a theory about the origin of words. Nevertheless, his theory of Universal Grammar requires the use of words.

To see how Chomsky dealt with this problem, we have to return to his most recent model of Universal Grammar, the strong minimalist thesis (SMT), not so as to comment on his contribution to our understanding of grammar (a daunting task that I will leave for other, more qualified individuals), but instead to see how he finessed the problem of the origin of words. In their place, he postulated innate "computational atoms" of thought. Computational atoms are lexical entities, "word-like inasmuch as they are discrete units of meaning, but not words because they have no phonological structure and no physical form outside the brain."[33]

LANGUAGE AS A BIOLOGICAL ORGAN

How do computational atoms become words? Chomsky's answer is through "externalization," one of the processes of the SMT. To see how that works, we have to abandon our familiar view of language and accommodate ourselves to his distinction between *internal* and *external* language and his view that language is a biological (mental) organ that is "on a par with the visual or digestive or immune systems."[34] Internal language is innate. Its main function is to help us think. External language is what people use to communicate and to disseminate knowledge[35]. That, of course, is the normal meaning of language.

Why should we accept Chomsky's view of language as a biological organ? Chomsky argues that the distinction between internal and external language is necessary if we want to understand language in its pure form. It's a step that Chomsky likens to Galileo's abstraction of the motion of fallen bodies to eliminate extraneous factors that could influence their fall. To study the properties of internal language, Chomsky proposed an "ideal speaker-listener": someone who knows a language perfectly and is unaffected by such grammatically irrelevant conditions as memory limitations, distractions, shifts of attention, and so on.[36]

That justification assumes that we know much more about language than we do. Physicists have little trouble defining and measuring their subject matter in exacting detail, gravity in particular. As Galileo noted, when measuring gravity in the real world, estimates of the rate of fall of a body may vary because of differential resistance to friction from air. Galileo eliminated that source of variability by calculating the rate at which an idealized body falls.

The study of language, however, is vastly different from the study of physics. Linguists have yet to agree on a definition of language or on the universality of Universal Grammar.[37] Without a concrete model of an ideal speaker-listener, it's not clear how to measure deviations from that ideal to control for grammatically irrelevant conditions. Another feature of language that doesn't lend itself to the fallen-body model is how it develops from infancy to adulthood. Chomsky's ideal-speaker listener concept arguably applies to an adult, but it's not applicable to the first words or sentences of an infant.

Another seldom-recognized limitation of the "language as a biological object" approach is that language requires two people and the biological approach recognizes only one. It is certainly not the case, as Chomsky asserts, that "the language organ grows like any other body organ."[38] As I made clear in chapter 4, children learn to talk only through the intervention of their caretakers, who guide them through the stages of intersubjectivity and joint attention. There is no reason to suppose that language would mature in an infant who was raised alone.

As compared to earlier versions of Universal Grammar, Chomsky argues that the SMT is simple enough to have resulted from a single mutation. That allowed him to reduce the evolution of language to a three-part system. First is a simple internal computational system that uses *merge* to build hierarchically structured thoughts. That system interfaces with two others: a sensorimotor system for externalizing the structures produced by the computational system, and a conceptual system for inference, interpretation, planning, and the organization of action. Externalization converts internal thoughts into overt expressions: for example, into speech or, in the case of individuals who are hard of hearing, into sign language. The internal conceptual computation system draws on various cognitive abilities, such as short- and long-term memory, attention, priming, and the like.

Internal language is generated by a mental computational system in which *merge* combines individual "computational atoms" into complex thoughts.[39] In turn, they can be combined computationally. For example, *merge* can combine the computational atoms *read* and *books*, into a new syntactic object, *read books*. *Merge* can then combine that object with a new one to produce *the man read books*. Embedded sentences provide another example: *The dog barked* can become *The dog, the man chased, barked*; or *The dog, the man chased, who spilled some coffee, barked*; and so on. This description of *merge* applies to current languages. Chomsky has yet to specify how it would have applied before externalization evolved; that is, before there were words.

Much has been written about the grammatical power of the SMT. For our purposes, however, I discuss only its treatment of words. That is the topic of the next few sections.

THINKING AND COMMUNICATION
BEFORE PROMETHEUS

Remarkably, Chomsky argued that external language (that is, the expression of words) evolved years *after* internal language appeared. That means that our ancestors were engaged in thinking privately, years before they ever used words!

The mutation that gave rise to internal language (and, indirectly, words) occurred about 80,000 years ago in some ancestor whom Chomsky dubbed Prometheus.[40] Prometheus benefited by being able to think, using computational atoms to do so. But because externalization had yet to evolve, Prometheus could not express his thoughts overtly. Eventually, Prometheus's mutant genes were passed to surviving children who thrived. Only then did human language, in its myriad of forms, emerge through externalization. As Chomsky wrote,

> Within some small group from which we are all descended, a rewiring of the brain took place in some individual, call him *Prometheus*, yielding the operation of unbounded Merge, applying to concepts with intricate (and little understood) properties . . . . Prometheus's language provides him with an infinite array of structured expressions . . . . Prometheus had many advantages: capacities for complex thought, planning, interpretation, and so on. The capacity would then be transmitted to offspring . . . . At that stage, there would be an advantage to externalization, so the capacity might come to be linked as a secondary process to the sensory motor system for externalization and interaction, including communication . . . .[41]

There are some obvious problems with the Prometheus proposal. The most serious is the requirement for a Mrs. Prometheus to transmit the mutation in question. The appearance of two "hopeful monsters" would be quite a stretch. Even if that improbable event occurred, the communicative function of language was required. How else could the advantages of the language of thought "take over a small group"?

There is a related question: What is the nature of thinking in ancestors who lived before Prometheus? Chomsky does not explain how thinking in those ancestors differed from well-known examples of animal cognition (e.g., categorization[42] and serial learning[43]). That question is itself the flip side of another question: What is the origin of computational atoms? They were supposedly innate, but how would that explain computational atoms such as *book*, *read*, and the like in the preceding example? That would require exemplars of innate concepts waiting to be expressed under the right circumstances. As I discuss in the next section, Chomsky advanced precisely that view in his explanation of reference.

THINKING AND COMMUNICATION
AFTER PROMETHEUS

Chomsky maintains that the main function of language, as it is currently practiced, is thinking rather than communication:

> The modern conception—that communication is the "function" of language (whatever exactly that is supposed to mean)—probably derived from the mistaken belief that language somehow *must* have evolved from animal communication [whose only function was communication] . . . . One might speculate that the modern conception also

derives from lingering behaviorist tendencies, which have little merit . . . . The evidence available appears to favor the traditional view that language is fundamentally a system of thought [italics in original].[44]

Chomsky justifies his claim that the main function of language is to help us think by asserting that "it takes an enormous act of will to keep from talking to oneself in every waking moment."[45] That view makes one wonder whether it would be true of a one-year-old infant about to utter her first words. If not, as is obviously the case, at what point does one begin to "talk . . . to oneself in every waking moment"? Talking to oneself occurs in adults, but before a child can think autonomously, there is ample evidence that she had to learn to think by interacting with others.

To support his view that thought is the main function of language, Chomsky appeals to "the traditional view." As evidence, he cites statements by two Nobel laureates whose careers were in biology. Salvador Luria claimed that communicative needs would not have provided "any great selective pressure to produce a system such as language." Françoise Jacob argued that "the role of language as a communication system between individuals would've come about only secondarily . . . . The quality of language that makes it unique does not seem to be so much its role in communicating directives for action" but rather "its role in symbolizing, in the invoking [of] cognitive images."[46] That may be true of adults, but as noted earlier, autonomous thought doesn't come automatically. It requires training in how to think.

Equally puzzling is Chomsky's unwavering resistance to the communicative function of language in humans. He seems to have a blind spot regarding conversation, an activity that is uniquely human *and* communicative. There is still much to learn about the evolution of conversation, but the *fact* of conversation, which unquestionably did evolve, cannot be dismissed as a "belief."

As I discuss later, Chomsky's blind spot for conversation is also consistent with a general confusion about the meaning of *word*. The dictionary definition of *word* is a "speech sound or series of speech sounds that symbolizes and communicates a meaning usually without being divisible into smaller units capable of independent use."[47] To be meaningful, the word must be understood by a speaker and a listener to refer to a particular object or event. That's the equivalent of saying that words must be declarative or conversational, but no less an authority than Chomsky referred to Nim's signs as words. In discussing his signing ability, Chomsky explained why Nim's "two-word signs" were not syntactical,[48] ignoring the fact that none of Nim's signs were declaratives.

LEARNING CONCEPTS

Although the origin of words is "entirely obscure," they are somehow related to computational atoms, their innate "mental" counterpart. That creates a dilemma. How do computational atoms acquire word-like status? Chomsky never answered that question, but, as mentioned earlier, he implies that it has something to do with the similarity between languages and the immune system.

Consider Chomsky's example about how a person learns to name two modern entities: a *carburetor* and a *bureaucrat*. He acknowledged that those names would have to be learned communicatively (that is, conversationally). He nevertheless argues that the SMT can anticipate novel meanings. Just as the immune system can react to previously unknown, unanticipated viruses or other dangerous antigens, the lexicon of computational atoms can anticipate novel meanings. In effect, Chomsky is saying that humans evolved with an innate set of concepts that allow us to name objects that nature never anticipated, just as the immune system can resist novel pathogens.

> [T]here is good reason to suppose that [this] argument is at least in substantial measure correct even for such words as *carburetor* and *bureaucrat*, which, in fact pose this familiar problem of poverty of stimulus . . . . However surprising the conclusion may be that nature has provided us with an *innate stock of concepts*, and that the child's task is to discover their labels, the empirical facts appear to leave open few other possibilities [italics added].[49]

Chomsky provided no reference to the "empirical facts" that would explain the origin of concepts such as *carburetor* or *bureaucrat*. His construct of "an innate stock of concepts" seems to have the same miraculous force as the mutation that created grammar and computational atoms. We are still left with a double mystery about the origin of words and computational atoms.

THE ORIGIN OF REFERENCE

Chomsky has yet to acknowledge well-documented explanations of how children learn that words refer to particular objects,
a clear instance of communication (cf. chapter 4). Consider,
for example, his distinction between "mind-dependent" and
"mind-independent" entities. A brief translation is in order.
Mind-independent entities refers to the content of animal
minds, entities about which animals supposedly cannot think.
Mind-dependent entities refers to thoughts about objects that
are not physically present, objects about which humans have
to think.

> [Mind-dependent thought] is . . . uniquely human (and
> differs from) mind-independent entities that appear
> characteristic of animal communication . . . . For human
> language and thought, it seems, there is no *reference* relation
> in the sense of Frege, Peirce, Tarski, Quine and contempo-
> rary philosophy of language in mind. What we understand
> to be a river, a person, a tree, water, and so on, consistently
> turns out to be a creation of what 17th-century investigators
> called the human "cognoscitive powers," which provided us
> with rich means to refer to the outside world from intri-
> cate perspectives . . . . [I]t is only by means of the inward
> ideas produced by its "innate cognoscitive power" that the
> mind is able to know and understand all external individ-
> ual things . . . . The objects of thought constructed by the
> cognoscitive powers cannot be reduced to a peculiar nature

of belonging to the thing we are talking about [italics in original].[50]

Here again, Chomsky assumes an adult's knowledge of language. "What we understand to be a river . . ." is a description of an *adult's* thoughts about a river, reflections that are colored by earlier conversations about rivers, nature, and so on. Until a child learns to name a river, her initial experience of a river is like an animal's, what Chomsky refers to as mind-independent.

I don't question the development or distinction of mind-independent from mind-dependent concepts, but it is not clear how a "cognoscitive power" helps us to understand that process. As far as I'm aware, cognoscitive powers are not a recognized psychological phenomenon. Nor is the dictionary definition of *cognoscitive* helpful: "having the power of knowing." More problematic is the absence of any basis for Chomsky's claim that "there is no reference relation in the sense of Frege, Peirce, Tarski, Quine," a claim that ignores widely accepted evidence that joint attention helps children learn to name objects and thereby refer to them.[51]

It is also not clear what would be lost if we were to substitute "naming" for mind-independent thoughts and "categorization" for mind-dependent thoughts.[52] It has been well established that animals can learn categories. For example, in an experiment with monkeys, I showed that they could be trained to differentiate novel photographs of flowers from novel photographs of trees from novel photographs of people, and so on, on a first-trial basis.[53] They could not do so unless they had already formed the relevant categories "flower," "trees," "people," and the like. As also shown in the case of monkeys that develop serial expertise (see chapter 2), animals can think. They do that without language, that is, without naming things. If they could name things, they would have mind-dependent thoughts. Without names they are limited to mind-independent thoughts.

A FEAR OF BEHAVIOR?

Chomsky's justification for studying internal language is that it is not contaminated by limitations of memory, attention, and the like. Given our limited understanding of language, that concern is premature. A more obvious reason for Chomsky's preoccupation with internal language is his fear of behavior, the bête noir of generative grammar. In rationalizing SMT, Chomsky observed that the "modern conception [of language as communication] . . . derives from lingering behaviorist tendencies which have little merit."[54]

To understand Chomsky's fear of behavior, we have to return to his critique of *Verbal Behavior*,[55] a critique for which he is justly famous. There were two aspects to that critique. One showed the fatal weaknesses of Skinner's explanation of sentences. As I mentioned in chapter 1, those arguments are sound and have stood the test of time. But what about words? Words are unequivocally behavior. Although Chomsky rejected Skinner's account of sentences, he presented no arguments against Skinner's behavioral account of the origin of words.

In the first chapter of *Verbal Behavior*, Skinner presented his brilliant insight concerning what is special about language. He defines *language* as "behavior reinforced through the mediation of other persons." In chapter 1, I gave some examples of word-learning from *Verbal Behavior*. In the presence of a discriminative stimulus (e.g., a glass of water), the spoken or written word *water*, spoken or written H_2O, the child's response of *water* is acknowledged by the parent saying, *right, yes, that's water*, or something similar.

Skinner's examples are, of course, dated. He conceived of word-learning more than thirty years before joint attention was discovered (see chapter 4). Skinner nevertheless put his finger on the central fact about word-learning. Initially, word-learning is conversational, a fact that Chomsky ignored. Indeed, babies begin to communicate with their mothers preverbally by their fourth month.[56] Words are learned conversationally at the beginning of an infant's first year. As far as I'm aware, there's no better account of word-learning.

If Chomsky tried to describe word-learning by children, he would have to come to grips with an infant's unique preverbal path to language and the conditions under which the infant learns to utter her first words. Instead, he ignored word-learning and professed that SMT is a solution. The only benefit of postulating innate computational atoms is that they are not tainted by the conversational process that is necessary for the acquisition of words. Only then could Chomsky claim that thinking is the main function of language and that its role in communication is secondary. As noted earlier, however, once Chomsky recognizes that words exist, he has to explain how they can be expressed. That can only happen conversationally.

WALLACE'S PROBLEM REVISITED

Earlier, we saw that Wallace framed the basic question about the evolution of language by asking how the addition of language would enhance the survival of a species. There have been many attempts to answer that question, but as we saw earlier, none of them explained how words would have increased the survival of a species—a crucial condition for arguing for natural selection.

Bickerton's theory seems to provide the best answer. As is generally true of theories about evolution, the only evidence is circumstantial, but that is true of all theories of evolution, whether of the universe or of language. The goal of such theories is simply to stimulate discussion about their validity by ruling out obvious and less obvious flaws.

That seems to be the motivation of Chomsky's book (with Berwick), *Why Only Us*, a title which, curiously, lacked a question mark.[57] That book, the subtitle of which is *Language and Evolution*, asserted that language resulted from a relatively recent mutation.[58] A few years earlier, Berwick and Chomsky were among the coauthors of an article in which they lamented the fact that "the most fundamental questions about the origins and evolution of our linguistic capacity remain as mysterious as ever."[59] I don't agree with their conclusion but I'm sympathetic to its view that there is much to learn about the evolution of language, even though what we have are at best educated guesses. Just as theories of the evolution of the universe have advanced our knowledge of astronomy, theories of the evolution of language will advance our knowledge of humans' place in nature.

Whatever the starting point of a theory about the evolution of language, it must try to answer Wallace's question. As mentioned earlier, Chomsky's theory fails that test because there is no reason to suppose that our ancestors needed *merge* to survive.[60] Bickerton's theory has more promise than Chomsky's because it is supported by circumstantial evidence: the relative size of *Homo erectus*'s brain, a dry environment, its need for meat, and the distribution of bite and cut marks on bones of large fauna. As discussed earlier in this chapter, those circumstances may have provided the impetus for creating the first words, which Bickerton argued referred to remote sources of meat.

By contrast, the only evidence that Chomsky cites to support his claim about a mutation is weak. There is a large gap between animal communication and language and the appearance of what Chomsky described as the oldest artifacts of art shortly after the purported mutation that produced language.

Let's first consider the gap between animal communication and language. As I've argued throughout this book, that gap would be easier to bridge by breaking down the evolution of language into two smaller steps: the evolution of words and the evolution of grammar. Most linguists agree that words evolved before grammar and support the view that proto-language existed before grammatical language emerged.[61]

The gap between animal communication and language could be further reduced by relaxing Chomsky's claim that grammar appeared abruptly. He took that position early in his career because he thought that a partial grammar for humans would be as useless as partial wings for birds. That claim always struck me as dubious. Just as a bird with a partial wing could still escape from a predator more readily than one without any wings, a person with a partial grammar (say, one who could only produce the present tense) would have an advantage over a person who lacked any grammar.

Chomsky's view of the abrupt emergence of grammar has been challenged by psycholinguists who claim that the emergence occurred in stages.[62] Although there is no evidence that can adjudicate between the gradual and abrupt appearance of grammar, there doesn't appear to be any *a priori* reason to argue that grammar did not evolve in steps of increasing complexity, just as it does in children.

Turning to the relation between a mutation that created grammar and prehistoric art, recent evidence casts doubt on the validity of Chomsky's assumption that prehistoric art first appeared about 80,000 years ago. As archaeologists uncover more examples, the age of prehistoric art keeps receding. When Chomsky claimed that the mutation that created language occurred about 80,000 years ago, the oldest example of prehistoric art created by a *Homo sapiens* was an ocher drawing on a flake of silcrete in the Blombos cave of South Africa.[63] In 2018, archeologists discovered evidence of Neanderthal cave drawings in a Spanish cave.[64] *Homo neanderthalis* appeared at least 400,000 years ago. If that report is validated, and if one accepts the premise that the ability to produce art is evidence of symbolic language,[65] it will show that at least one other species had language, and that language and art evolved much earlier than Chomsky supposed.

THE FUTURE OF CHOMSKY'S
APPROACH TO LANGUAGE

The reader will recall that the starting point of this book was Nim Chimpsky's failure to produce sentences in American Sign Language. That failure confirmed Chomsky's view that language was uniquely human, but it obscured the fact that the real reason for the failure was Nim's inability to learn words. Nim's failure raised a more specific question: When and how did words emerge? Chomsky's inability to answer that question raises doubts about the viability of his effort to understand the evolution of language. Just the same, I believe that his search for a Universal Grammar that unites all languages will ultimately succeed, but not his hypothesis about a genetic mutation.

Chomsky's quest for a Universal Grammar is reminiscent of scientists' search for the periodic table, the system that shows the relation of all of the elements. Until Mendeleev created the periodic table, chemists couldn't understand how particular elements combine to form compounds. Linguists, including Chomsky, are still struggling to describe the features of words upon which *merge* operates,[66] let alone determine rules for combining them.

A comparison of the search for Universal Grammar and the periodic table obscures two reasons why the search for

Universal Grammar is much more difficult. The first is our lack of understanding of the origin of words. Relatively speaking, the origin of the elements that comprise the periodic table is a simple problem. The second is that the origin of words, both phylogenetically and ontogenetically, requires an understanding of the transmission of knowledge between two individuals, a problem that still baffles psychologists. The nature of that process has only recently come into focus.[67]

Recognizing the need for a Universal Grammar is a brilliant insight. However, it will not be discovered without understanding the origin of words and how words set the stage for the evolution of grammar. With that caveat in mind, I recommend that the search for a Universal Grammar should:

(1) Acknowledge and integrate the discoveries of developmental psychologists about the importance of joint attention for the initial production of words. In this connection, I can envision the incorporation of intersubjectivity and joint attention into the language acquisition device concept.

(2) Recognize that words are just as uniquely human as grammar. They should be included in what Chomsky and his colleagues refer to as the "Faculty of Language in the Narrow Sense."[68]

(3) Recognize that a Universal Grammar can emerge gradually. There is no a priori reason to argue that its universal features had to emerge simultaneously.

None of this will detract from the importance of Chomsky's concept of Universal Grammar, for which there is no viable alternative. If anything, it should place Universal Grammar on a firmer ground that would only increase its influence. Taken together, these steps should simplify what Christiansen and Kirby have called "the hardest problem of science": the evolution of language.[69]

■ ■ ■

Summary: The goal of ape language experiments was to obtain evidence to support Darwin's view of a continuous progression from animal communication to language. At the beginning of his career, Chomsky attacked that idea by arguing that language could not have evolved from animal communication because the gap between animal communication and language was too large to have been filled by the small steps that natural selection would require. Instead, he argued that language is somehow produced by a Universal Grammar, an innate neural mechanism that gives rise to the sentences of any of the world's languages.

More recently, Chomsky changed his position about the theory of evolution. Universal grammar was simplified and reduced to a single operation, *merge*. Its simplicity motivated Chomsky to hypothesize that Universal Grammar was the result of a mutation that occurred about 80,000 years ago. That would explain the evolution of language, not by natural selection, but by a small, genetically induced rewiring of the brain. That hypothesis, however, said nothing about the origin of words, without which a grammar could not work.

In 1869, Wallace, the codeveloper of the theory of evolution, argued that human language far exceeds what is needed for humans to survive. Wallace argued that language could not be a product of evolution because he couldn't see how it could be explained by natural selection.

If Wallace were alive today, he would probably recognize that theories of the evolution of language have to address two issues: the transition from animal communication to words; and, ultimately, how grammar evolved. He might also have recognized that an answer to the first question would address his concern about the role of natural selection in the evolution of language.

Bickerton proposed an answer to Wallace's problem. He noted an unusual set of circumstances that might have contributed to the use of words by *Homo erectus*: its relatively large brain (the largest of any hominin) and its need for a diet that was rich in calories. Meat was the best source of those calories. Although *Homo erectus* lacked the ability to hunt large animals, they could use stone tools to cut through a dead animal's skin to obtain meat.

After members of a group of *Homo erectus* encountered a dead animal, they needed to convey that information to other members of their group to obtain help in scavenging that animal for meat and warding off other animals that may have had a similar goal. To solicit help from colleagues who were too far away to see the dead animal, a scout had to use a new form of communication to convey the relevant information. Whatever form it took (gestures, mime, sound, or some combination thereof), that communication had to use arbitrary symbols that referred to an object that others could not see. Those symbols, which *Homo erectus* had to invent, distinguished that species' communication from animal communication.

That type of interaction also required a much greater degree of cooperation than that observed in apes. According to Hrdy, *Homo erectus* acquired that ability by virtue of being the first hominin to engage in cooperative breeding, a practice whereby a mother allows other members of her group to help rear her infants.

Although Chomsky's theory of language cannot account for words *per se*, he proposed "computational atoms" as word-like units of meaning. According to Chomsky, computational atoms resulted from the mutation that produced language and are important because they are a vehicle of internal thought. They are also the basis for the primary function of language, which for Chomsky is thinking, not communication. Words as we know them evolved later as "externalized" units of meaning.

Chomsky offered no support for the claim that thinking is the main function of language. Indeed, it is contradicted by evidence that children do not think until after they learn to talk. His theory of language also fails to consider empirical evidence of intersubjectivity and joint attention and their role in the communicative function of language.

Chomsky revolutionized linguistics by introducing the concept of Universal Grammar, which was a significant advance over the structural descriptions of grammar that dominated linguistics during the first half of the twentieth century. However, his theory of Universal Grammar did little to explain the origin of words or their function. Like most theories of grammar, they have diverted attention from the first step in the evolution of language: the shift from animal communication to the use of words.

EPILOGUE

PROJECT NIM, A DOCUMENTARY FILM

AS A COMPARATIVE psychologist, I study the behavior of various animal species. Until Project Nim, I was best known for my work with pigeons and monkeys. Early in my career, as a student of B. F. Skinner at Harvard University, I showed that pigeons could learn to discriminate colors and geometric forms without making any errors.[1] That demonstrated, for the first time, that an animal does not need to rely on trial and error to differentiate arbitrary stimuli. The logic I followed was similar to that used by Skinner when he developed the teaching machine: start with a simple problem and gradually increase its difficulty.[2] Computerized versions of the teaching machine have been widely used to train people to learn various kinds of technical facts with a minimum number of errors.

Many years later, I trained monkeys to produce arbitrary sequences containing seven or more items. The items (photos) were presented simultaneously on a touch-sensitive video monitor.[3]

The monkey's task was to touch them in the correct order, irrespective of their positions on the monitor. People execute similar sequences when they enter a password to operate an ATM, but people can encode such items linguistically. Because monkeys cannot, it's not clear how they remember such long sequences. To understand how monkeys could think without language, I devised experiments on animal cognition, many of which I am still conducting.[4]

The results of my experiments on pigeons and monkeys have been well received and replicated by other psychologists. They have also motivated much additional research. It is therefore ironic that I am less well known for those experiments than I am for a single experiment that I performed on a single subject, in which I tried to teach a chimpanzee, Nim Chimpsky, some rudiments of a human language.[5] The results were negative and controversial. Unlike my other research, no one has attempted to replicate it. Particularly in psychology, negative results rarely cause much of a stir. What was different about Project Nim? Experiments on the evolution of language are rare and negative results are even rarer.

Project Nim was motivated by a debate about the nature of language between Skinner and Noam Chomsky: an argument between the leading behaviorist of the time and a young linguist who was one of the founders of the so-called "cognitive revolution."[6] Shortly after Skinner published *Verbal Behavior*,[7] a book in which he argued that language can be reduced to sequences of conditioned behavior, Chomsky wrote a scathing review in which he showed why Skinner was wrong. He also claimed that language was innate and uniquely human.

Chomsky further argued that the ability to create sentences was the basic feature of language.

The reaction to Chomsky's claim was predictable. Some behaviorists (myself included) began projects in which we tried to show that language was not uniquely human by teaching chimpanzees to use American Sign Language.[8] Project Nim was an example.[9]

In 2010, James Marsh, a well-known filmmaker who received an Academy Award in 2008 for his documentary *Man on Wire*, offered to make a full-length documentary about Project Nim. Having participated in other documentaries about Nim,[10] I welcomed the opportunity to inform the public about the rollercoaster experience I had while teaching Nim to use American Sign Language. At the time (in the 1970s), I thought I had shown that Nim produced short sentences, only to discover later that language was not needed to explain those sequences. I wanted to convey the exhilaration of thinking that our closest living ancestor had learned to communicate in a human language, and the disappointment I felt when I realized that my results were negative.

Although I discussed these and related issues with Marsh, his documentary ignored the science that motivated Project Nim and the implications of its negative results. Because *Project Nim* never mentioned Chomsky, or his significance as a linguist, it never explained the origin of Nim's surname. Marsh also failed to mention that the goal of Project Nim was to train Nim to produce sentences, the type of evidence that was needed to achieve that goal, why it seemed that Nim had produced sentences, or what made me change my mind.

Curiously, Marsh did include in *Project Nim* a videotape I made that showed why my results were negative. Nim can be seen asking his teacher, in sign language, to play with a cat. The videotape also shows that the teacher prompted Nim by signing "cat," "you," "Nim," and other signs. *Project Nim* made no mention of the teacher's cues and how they contributed to my negative conclusions about Nim's ability to create sentences.

Marsh also ignored the fact that Project Nim was a replication of Allan and Beatrice Gardner's attempt to train the chimpanzee Washoe to learn sign language.[11] That's a shame, because science thrives on scientists' efforts to replicate the results of other researchers and to make sense of negative results.

The history of science provides many famous examples. The modern theory of combustion is based on negative results from experiments that disproved the existence of phlogiston. Antoine Lavoisier showed that oxygen from the air was the key ingredient. Albert Einstein developed the theory of relativity in response to the negative results of an experiment showing that light could travel through the universe without ether. Albert Michelson was awarded a Nobel Prize for that experiment.

Psychology is too young and too complicated a science to have generated negative results that are significant enough to motivate new theories. The negative results of Project Nim are nonetheless important because of their implications about the nature of language and its evolution. Most significantly, Project Nim revealed why the evolution of language was too broad a question and why it had to be broken down into smaller steps: how did words evolve and, given the evolution of words, how did grammar evolve?

I discussed these topics during lengthy on-camera interviews with Marsh. I explained that Nim was smart enough to learn how to produce signs to obtain rewards, but that videotapes revealed that his teachers had unwittingly cued most of his signs. Analyses of films of Washoe signing with her teachers showed a similar relation. None of these comments were included in *Project Nim*. Indeed, the only reference to the negative results I obtained was Marsh's innuendo that I returned Nim to the primate colony in which he was born as punishment for his failing to learn sign language! Indeed, after seeing *Project Nim*, many viewers sent me hate mail accusing me of being a monster who ruined Nim's life.

That was just one of many falsehoods presented in *Project Nim*. At the time, I thought I had enough evidence to justify my initial impression that Nim could produce sentences, and I had almost published an article saying so. It took me a year to discover that the evidence of grammatical regularities I had collected were an artifact of Nim's ability to imitate his teachers' signs. When I returned Nim to his birthplace in Oklahoma, I thought he had learned language.

Because Marsh didn't mention that Project Nim was a replication of other ape language projects, his film didn't include criticisms by other investigators[12] or my proposal for answering those critics.[13] An unedited videotape of an ape signing with a teacher is all that would be needed: it would reveal the extent to which the ape was prompted to sign and the extent to which it signed spontaneously.

Project Nim, which was released in 2012, was mainly an ad hominem attack on me that consisted of interviews with

Nim's teachers. Their memories of the project differed significantly from mine. But this is not the place for a point-by-point rebuttal.[14] What disappointed me most about the film was its complete failure to present the scientific background of the work we did with Nim and its theoretical significance.

Because regulations by the National Institutes of Health now prohibit research on chimpanzees, it would be virtually impossible to launch a new ape language project. Even if one were possible, investigators should keep in mind comments made by Thomas Sebeok, a noted linguist, who reviewed the various ape language experiments of the 1970s. Sebeok remarked that those "experiments . . . divide into three groups: one, outright fraud; two, self-deception; three, those conducted by Terrace. The largest class by far is the middle one."[15] Because Marsh made it seem that Nim learned sign language, *Project Nim* fell in the middle category.

There is, however, one thing about which Marsh and I can agree. Viewers of *Project Nim* would have to close their eyes not to appreciate Nim's lovable personality and his remarkable intelligence and empathy. Even though we didn't succeed in communicating with him linguistically, we all thought of him as a fascinating living relative from the tree of evolution. He deserves a place in history for sharing himself and his abilities in the pursuit of what it means to be human and for helping us to understand what he and his descendants are and are not.

NOTES

PREFACE

1. Terrace (2012).
2. Chomsky (1959); Skinner (1957).
3. Metcalfe & Terrace (2013).
4. Studdert-Kennedy & Terrace (2017).

PROLOGUE

1. Terrace (1979); Terrace, Petitto, et al. (1979).
2. Berwick & Chomsky (2016).
3. Chomsky (1959).
4. R. A. Gardner & B. T. Gardner (1969); D. Premack (1971).
5. Hayes & Hayes (1951); Kellogg & Kellogg (1933); Kohts (1935).
6. Stokoe, Casterline, et al. (1965).
7. D. Premack (1976).
8. Rumbaugh, Gill, & von Glasersfeld (1973).
9. Brown (1970).
10. Kaye (2014).
11. L. M. Bloom (1973).
12. R. A. Gardner & B. T. Gardner (1973); Schroeder (1978).
13. Hauser, Yang, et al. (2014).

1. NUMBERLESS GRADATIONS

1. Darwin (1859).
2. Tennyson (1850).
3. Huxley (1887), 170.
4. Dobzhansky (1973).
5. Darwin (1872), 35.
6. Müller (1862).
7. Kenneally (2007).
8. Shively (1985).
9. Wyatt (2015); Hadhazy (2012).
10. Stoddard & Markham (2008); Wong & Hopkins (2007).
11. Langmore (1998); Fortune, Rodríguez, et al. (2011).
12. Aguilar, Fonseca, & Biesmeijerb (2005).
13. Cheney & Seyfarth (1992).
14. De Waal (2011).
15. Spencer (1886).
16. Köhler (1925).
17. Birch (1945).
18. Pavlov (1927).
19. Skinner (1938).
20. Skinner (1959).
21. Skinner (1957)
22. Skinner (1960).
23. Herrnstein (1985).
24. Harlow (1949).
25. Skinner (1959).
26. Sundberg & Michael (2001).
27. Copeland & Hall (1976); Luiselli, Putnam, et al. (2005).
28. Skinner (1957).
29. Chomsky (1959).
30. Pinker (1994).
31. Chomsky (1980).
32. Chomsky (1965).
33. Chomsky (1972).
34. Chomsky (1957).
35. Chomsky (1972).
36. Darwin (1859), 143.

37. Darwin (1859), 211.
38. Matticus78 (2006).
39. Darwin (1859), 508.
40. R. A. Gardner & B. T. Gardner (1969); D. Premack (1976); Rumbaugh (1977); Terrace, Petitto, et al. (1979).
41. Hayes (1951); Kellogg & Kellogg (1933).
42. Lieberman (1975).
43. Yerkes (1925).
44. Rosenberg & Trevathan (2002).
45. Bowlby (1980), 90; Hinde (1987); Hinde & Stevenson-Hinde (1987).
46. Jaffe & Anderson (1979).
47. Trevarthen & Aitken (2001); Trevarthen (1998).
48. Scaife & Bruner (1975); Bruner (1983).
49. Jaffe, Beebe, et al. (2001); Beebe, Stern, & Jaffe (1979).
50. Lavelli, & Fogel (2002), 288.
51. Beebe, Messinger, et al. (2016).
52. Quine (1960).
53. Wilkes-Gibbs & Clark (1992).
54. Carpenter & Call (2013).
55. Morales, Mundy, et al. (2000); Mundy & Newell (2007).

2. APE LANGUAGE

1. R. A. Gardner & B. T. Gardner (1969); D. Premack (1976); Rumbaugh (1977); Terrace, Petitto, et al. (1979).
2. R. K. Thompson & Herman (1977); Schusterman & Kastak (1998).
3. D. Premack (1976).
4. Kellogg & Kellogg (1933).
5. Hayes & Hayes (1951).
6. Yerkes (1925), 180.
7. R. A. Gardner & B. T. Gardner (1969).
8. Brown (1970).
9. Brown (1973).
10. Fouts (1975).
11. In ASL, the beginning and the end of a phrase are communicated by the raising and the lowering of the hands in front of the speaker's body. Fouts's diary omitted any mention of this aspect of ASL.

12. Terrace, Petitto, et al. (1979).
13. Terrace (1979).
14. Terrace (1979).
15. Terrace (1979).
16. Pfungst (1911).
17. R. A. Gardner & B. T. Gardner (1973).
18. Schroeder (1977).
19. Yang (2013).
20. Patterson (1981), 86–87; Gardner, B. T. (1981); Ristau & Robbins, (1982).
21. Terrace (1979).
22. Terrace (1985).
23. D. Premack (1976).
24. Figure adapted from A. J. Premack & D. Premack (1972).
25. Figure adapted from http://www2.gsu.edu/~wwwlrc/Media/Images/LAC /lexkb.JPG.
26. D. Premack (1976).
27. Rumbaugh, Gill, & von Glasersfeld (1973).
28. C. R. Thompson & Church (1980).
29. Savage-Rumbaugh, Rumbaugh, & Boysen (1978).
30. Savage-Rumbaugh, Rumbaugh, et al. (1980).
31. Jensen, Altschul, & Terrace (2013).
32. Savage-Rumbaugh, Murphy, et al. (1993).
33. The actual size of Kanzi's vocabulary for understanding spoken English or of lexigrams has not been documented.
34. Savage-Rumbaugh, Murphy, et al. (1993).
35. Savage-Rumbaugh, Murphy, et al. (1993), 98.
36. Savage-Rumbaugh, Murphy, et al. (1993), 130.
37. Savage-Rumbaugh, Murphy, et al. (1993), 119.
38. Wynne (2007).
39. Savage-Rumbaugh, Murphy, et al. (1993), 40.
40. Savage-Rumbaugh (1994).
41. Kaminski, Call, & Fischer (2004).
42. Markson & Bloom (1997).
43. Pilley & Reid (2011).
44. Because Chaser handled each object with her mouth, each object was frequently washed to eliminate acquired odors, thus the necessity of indelible marks.
45. Rumbaugh (1977).

46. Terrace (2005).
47. Savage-Rumbaugh, Rumbaugh, et al. (1980).
48. Bates (1979); Bates, Bretherton, et al. (1983).

3. RECENT HUMAN ANCESTORS AND THE POSSIBLE ORIGIN OF WORDS

1. Berwick & Chomsky (2016).
2. Mayr (2001).
3. Fu, Li, et al. (2014); Green, Malaspinas, et al. (2008); Sánchez-Quinto, Botigué, et al (2012).
4. Slon, Mafessoni, et al. (2018).
5. Folinsbee, Lipson, & Reynolds (1956); Walter (1997).
6. Coppens (1994).
7. Senut (2006).
8. Reader (2011).
9. McBrearty & Jablonski (2005).
10. Shreeve (1996).
11. Demenocal (2004); Potts (2012).
12. Cullen, deMenocal, et al. (2000); Guatelli-Steinberg (2003); Teaford & Ungar (2000); Maslin, Pancost, et al. (2012).
13. Potts (1996).
14. In mammals, kangaroos are an exception.
15. Larson (2009).
16. Simpson (1945).
17. Asfaw, White, et al. (1999).
18. Wood & Harrison (2011).
19. Antón (2003).
20. Dubois (1896, 1937).
21. Swisher III, Curtis, & Lewin (2001).
22. Antón (2003).
23. The oldest remains of *Homo erectus* are approximately 1.8 million years old, and there is evidence that they existed as recently as 300,000 years ago.
24. Rightmire (2013).
25. R. J. Blumenschine (2016) in a personal communication.
26. Bickerton (2014); Bickerton & Szathmáry (2011); Pobiner (2016).
27. Hrdy (2009).

28. Bickerton (2014); Dominguez-Rodrigo, Pickering, et al. (2005); Blumenschine (1987); Bunn, Kroll, et al. (1986); Pante, Scott, et al. (2015).

29. The results of a recent study of cut marks on animal fossils in areas of Africa infested by crocodiles (Sahle, El Zaatari, & White 2017) may require a reanalysis of cut- and bite-mark analyses performed at Olduvai sites (as reported by Dominguez-Rodrigo, Pickering, et al. 2005). In some sites in Ethiopia, which are about 4.2, 3.4, and 2.5 million years old, it was shown that marks on bones which were thought to have been caused by stone tools used by hominins were similar to marks left by crocodile bites. It has, however, yet to be shown if that equivalence is true of more recent cutmarks reported.

30. Blumenschine, Cavallo, & Capaldo (1994). Superficially, invertebrate species show evidence of displacement, e.g., ants (Wilson 1962) and bees (von Frisch, K. 1967.), but their communication of food locations are innate and unlearned.

31. Berwick & Chomsky (2016).

32. Everett (2012, 2017).

33. Jackendoff (1999); Progovac (2015).

34. Leland (2017).

4. BEFORE AN INFANT LEARNS TO SPEAK

1. Hobson (2002).

2. Berwick & Chomsky (2016), 56.

3. Chomsky (1965).

4. Chomsky (1975), 123.

5. Chomsky (1988), 173.

6. For ease of communication, I will refer to all of an infant's caretakers, here and elsewhere, as her "mother" even though it might be equally appropriate to refer to the infant's father, grandparents, siblings, friends or other caretakers.

7. DeSilva & Lesnik (2006).

8. Meltzoff & Moore (1977).

9. *See also* Nagy & Molnar (1994).

10. Meltzoff & Moore (1977).

11. Meltzoff & Moore (1998), 49.

12. Meltzoff & Moore (1989, 1994).

13. Imitation of the type described by Meltzoff (tongue protrusion and mouth opening) has been observed in other infant primates, e.g. chimpanzees and rhesus macaques in Bard, Myowa-Yamakoshi, et al. (2005); Tomonaga, Tanaka, et al. (2004). Non-human imitation does not, however, become a permanent part of those primates' repertoire.

14. Murray & Trevarthen (1985).

15. Tronick, Als, & Adamson (1979).

16. Beebe & Steele (2013).

17. Trevarthen (1977, 1980).

18. Stern (1971, 1985).

19. Beebe, Messinger, et al. (2016).

20. Beebe, Messinger, et al. (2016).

21. Bateson (1975, 1979).

22. Beebe, Alson, et al. (1988).

23. Beebe, Alson, et al. (1988); Jaffe, et al. (2001) performed an important review of rhythms of dialogue in infants.

24. P. Bloom (2000); Bickerton (2014); Bruner (1985); MacNamara (1972).

25. Bråten (1998); Bruner (1985); Dunham & Moore (1995); Scaife & Bruner (1975).

26. MacNamara (1972); D. Premack (1986); Quine (1960).

27. Clark (1996).

28. *See* Skinner (1957), for example.

29. Quine (1960).

30. Bruner (1983).

31. Wilkinson, Mandl, et al. (2010); Range & Virányi (2011); Schloegl, Kotrschal, & Bugnyar (2007); Kaminski, Riedel, et al. (2005); Teglas, Gergely, et al. (2012).

32. Tomasello, Carpenter, et al. (2005); Hobson (2002).

33. Hare, Call, & Tomasello (2001).

34. Hare, Call, et al. (2000).

35. Povinelli & Vonk (2003).

36. Warneken & Tomasello (2006); Warneken, Chen, & Tomasello (2006).

37. Carpenter & Call (2013).

38. Silk, Brosnan, et al. (2013).

39. Tomasello, Carpenter, et al. (2005).

40. Higgins (2016), 466.

41. Brooks & Meltzoff (2002).

42. Carpenter, Akhtar, & Tomasello (1998).

43. Morales, Mundy, et al. (2000); Mundy & Newell (2007).
44. Brooks & Meltzoff (2002); Brooks & Meltzoff (2008).
45. Powell (2010).
46. Mundy, Block, et al. (2007).
47. Tottenham, Hare, et al. (2010).
48. Eigsti, et al. (2011); Nelson, Zeanah, et al. (2007); Rutter (1998); Windsor, Glaze, et al. (2007).
49. Mundy, Sigman, & Kasari (1990).
50. Adamson, Bakeman, et al. (2017).
51. Charman (2003); Eigsti, et al. (2011); Mundy, Sigman, et al. (1990).

5. THE ORIGIN OF LANGUAGE, WORDS IN PARTICULAR

1. Hobaiter & Byrne (2014).
2. Saussure (1916, 1959).
3. Berwick & Chomsky (2016).
4. Like language, thinking has many meanings. I define it as being able to act on the memory of some object or event. This definition, which is true of both animal and human thinking (H. Gardner 1985), is not controversial in the case of human psychology. It reflects the obvious fact that humans can remember an object or event in their absence and act accordingly (that is, they can *re-present* to themselves some object or event). For behaviorists, this definition was controversial because behaviorism made no provisions for memory prior to the onset of experiments on animal cognition. Good introductions to that topic can be found in Roitblat, Bever, & Terrace (1984) and Wasserman & Zentall (2012).
5. Chomsky (1988).
6. Berwick & Chomsky (2016).
7. Berwick & Chomsky (2016), 1.
8. Berwick & Chomsky (2016).
9. Berwick & Chomsky (2016), 90.
10. Chomsky (1959).
11. Bickerton (2014), 91.
12. Gould (1977).
13. Wallace (1870).
14. Wallace (1870), 342.
15. Darwin (1869).

16. Bickerton (2014), 1.
17. Berwick & Chomsky (2016), 11.
18. Bickerton & Szathmáry (2011); Bickerton (2014).
19. Hockett & Altmann (1968).
20. Dominguez-Rodrigo, Pickering, et al. (2005); Blumenschine, Cavallo, & Capaldo (1994).
21. Bickerton (2009).
22. Arbib & Bickerton (2008); Wray (1998).
23. Hrdy (2009).
24. Burkhart, Hrdy, & van Schaik (2009); Hrdy (2009).
25. Greenfield (1991).
26. Deacon (1997).
27. Washburn & Lancaster (1968).
28. Falk (2004).
29. Dunbar (1998).
30. Mithen, Morley, et al. (2006).
31. Leland (2017).
32. In various articles, Chomsky estimated that language originated 70,000–200,000 years ago. For our purposes, how many years ago doesn't matter. What is critical is the sudden emergence of language.
33. Berwick & Chomsky (2016).
34. Berwick & Chomsky (2016), 56.
35. Chomsky metaphorically referred to the equivalent of an analogy between computational atoms and antibodies of the immune system. Just as an antibody can recognize a new antigen and manufacture an appropriate defense against that antigen, computational atoms can be modified by new objects to help a person find the appropriate word.
36. Chomsky (1959).
37. Evans & Levinson (2004)
38. Chomsky (1983).
39. Chomsky (1983), 70.
40. Chomsky (2013).
41. Chomsky (2005), 13.
42. Tanner, Jensen, et al. (2017).
43. Terrace (2005).
44. Berwick & Chomsky (2016), 102.
45. Ibid., 64.
46. Jacob (1982), as quoted in Berwick & Chomsky (2016), 81–82.

47. *Merriam-Webster Dictionary s.v.* "word," https://www.merriam-webster
.com/dictionary/word.
48. Berwick & Chomsky (2016),148.
49. Chomsky (2000), 65.
50. Berwick & Chomsky (2016), 84–85.
51. P. Bloom (2000); Tomasello (1999).
52. *See* Bickerton (2014, 79–80) for a discussion of naming and categorization.
53. Tanner, Jensen, et al. (2017).
54. Berwick & Chomsky (2016), 102.
55. Chomsky (1959).
56. Beebe, et al. (1988)
57. Berwick & Chomsky (2016).
58. *See* review by Studdert-Kennedy & Terrace (2017).
59. Hauser, Yang, et al. (2014). See reply by Terrace & Studdert-Kennedy
(2015).
60. *See also under* "Wallace's Problem," in this chapter."
61. For example, Bickerton (2014); Hurford (2007); Arbib & Bickerton (2008);
Wray (1998).
62. Progovac (2015); Heine & Kuteva (2007); Jackendoff (1999, 2002).
63. d'Errico, Henshilwood, et al. (2003).
64. Hoffmann, Standish, et al. (2018).
65. Miyagawa, Lesure, & Nóbrega (2018); Jones (February 23, 2018).
66. Evans & Levinson (2009); Dąbrowska (2015).
67. Schilbach, Timmermans, et al. (2013).
68. Hauser, Chomsky, & Fitch (2002).
69. Christiansen & Kirby (2003).

EPILOGUE

1. Terrace (1963a); Terrace (1963b); Terrace (1966).
2. Skinner (1959).
3. Terrace, Son, & Brannon (2003).
4. Roitblat, Bever, & Terrace (1984); Terrace, Son, & Brannon (2003);
Tanner, Jensen, et al. (2017).
5. Terrace, Petitto, et al. (1979).
6. H. Gardner (1985).
7. Skinner (1957).

8. Aside from the Gardners (R. A. Gardner & B. T. Gardner 1969), one of their students, Fouts (1975), directed a project whose goal was to teach a chimpanzee to use American Sign Language. Another method to avoid the articulatory bottleneck that prevents chimpanzees from communicating via a vocal language was to train them to use artificial languages composed of arbitrary visual symbols. Those experiments, D. Premack (1976) and Rumbaugh (1977), are discussed in Chapter 2.

9. Terrace (1979).

10. Harrar, WGBH, et al. (1984).

11. R. A. Gardner & B. T. Gardner (1969).

12. B. T. Gardner (1981); Miles (1983); Patterson (1981); Terrace (1981).

13. Terrace (1985).

14. A rebuttal of the claims made in *Project Nim* appears in the website originofwords.org (2019).

15. As quoted in Wade (1980).

REFERENCES

Adamson, L. B., Bakeman, R., Suma, K., & Robins, D. L. (2017). An expanded view of joint attention: skill, engagement, and language in typical development and autism. *Child Development*, 90(1), 1–18.

Aguilar, I., Fonseca, A., & Biesmeijerb, J. C. (2005). Recruitment and communication of food source location in three species of stingless bees (*Hymenoptera, Apidae, Meliponini*). *Apidologie*, 36(3), 313–324.

Alexeev, V. P. (1986). *The Origin of the Human Race*. Moscow, Progress Publishers.

Antón, S. (2003). Natural history of *Homo erectus. Yearbook of American Journal of Physical Anthropology*, 122, 126–170.

Antón, S. C., Potts, R., & Aiello, L. (2014). Human evolution. Evolution of early *Homo*: An integrated biological perspective. *Science*, 345(6192), 45–59.

Arbib, M. A., & Bickerton, D., eds. (2008). The emergence of protolanguage: Holophrasis vs. compositionality. [Special issue] *Interaction Studies*, 9(1).

Asfaw, B., White, T., Lovejoy, O., et al. (1999). *Australopithecus garhi*: A new species of early hominid from Ethiopia. *Science*, 284(5414), 629–635.

Bard, K., Myowa-Yamakoshi, M., Tomonaga, M., Tanaka, M., Costall, A., & Matsuzawa, T. (2005). Group differences in the mutual gaze of chimpanzees (*Pan troglodytes*). *Developmental Psychology*, 41(4), 616–624.

Bates, E. (1979). The emergence of symbols: Ontogeny and phylogeny. *Children's Language and Communication*, 12, 121–155.

Bates, E., Bretherton, I., Shore, C., & McNew, S. (1983). *Names, gestures, and objects: Symbolization in infancy and aphasia*. Hillsdale, NJ: Lawrence Erlbaum.

Bateson, M. C. (1975). Mother-infant exchanges: The epigenesis of conversational interaction. *Annals of the New York Academy of Sciences*, 263, 101–113.

——. (1979). The epigenesis of conversational interaction: A personal account of research development. In M. Bullowa (ed.), *Before speech: The beginning of interpersonal communication*, 63–78. Cambridge, UK: Cambridge University Press.

Beebe, B., Alson, D., Jaffe, J., Feldstein, S., & Crown, C. (1988). Vocal congruence in mother-infant play. *Journal of Psycholinguistic Research*, 17(3), 245–259.

Beebe, B., Cohen, P., & Lachmann, F. (2016). *The mother-infant interaction picture book: Origins of attachment*. New York: Norton.

Beebe, B., Messinger, D., Bahrick, L. E., Margolis, A., Buck, K., & Chen, H. (2016). A systems view of mother-infant face-to-face communication. *Developmental Psychology*, 32(4), 556–571.

Beebe, B., & Steele, M. (2013). How does microanalysis of mother-infant communication inform maternal sensitivity and infant attachment? *Attachment and Human Development*, 15(5–6), 583–602.

Beebe, B., Stern, D., & Jaffe, J. (1979). The kinesic rhythm of mother-infant interactions. In A. W. Siegman & S. Feldstein (eds.), *Of speech and time: Temporal patterns in interpersonal contexts*, 23–24. Hillsdale, NJ: Lawrence Erlbaum.

Berger, L. R., de Ruiter, D. J., Churchill, S. E., Schmid, P., Carlson, K. J., Dirks, P. H. G. M., & Kibii, J. M. (2010.) *Australopithecus sediba*: A New Species of Homo-Like Australopith from South Africa. *Science*, 328, 195–204.

Berger, L. R., Hawks, J., de Ruiter, D. J., Churchill, S. E., Schmid, P., Delezene, L. K. . . . & Zipfel, B. (2015). *Homo naledi*, a new species of the genus Homo from the Dinaledi Chamber, South Africa. *eLife*, 4, e09560.

Bermúdez de Castro; Arsuaga; Carbonell; Rosas; Martinez; Mosquera (1997). A hominid from the Lower Pleistocene of Atapuerca, Spain: possible ancestor to neandertals and modern humans. *Science*. 276 (5317): 1392–1395.

Bernard Wood (2011). *Wiley-Blackwell Encyclopedia of Human Evolution*. 2 Vols. New York: Wiley. 761–762.

Berwick, R. C., & Chomsky, N. (2016). *Why only us*. Cambridge, MA: MIT Press.

Bickerton, D. (2009). *Adam's tongue: How humans made language, how language made humans*. New York: Hill and Wang.

———. (2014). *More than nature needs: Language, mind, and evolution*. Cambridge, MA: Harvard University Press.

Bickerton, D., & Szathmáry, E. (2011). Confrontational scavenging as a possible source for language and cooperation. *Evolutionary Biology*, 11, 261.

Birch, H. (1945). The role of motivational factors in insightful problem-solving. *Journal of Comparative Psychology*, 38(5), 295–317.

Bloom, L. M. (1973). *One word at a time: The use of single word utterances before syntax*. The Hague: Mouton.

Bloom, P. (2000). *How children learn the meanings of words*. Cambridge, MA: MIT Press.

Blumenschine, R. J. (1987). Characteristics of an early hominid scavenging niche. *Current Anthropology*, 28(4), 383–404.

Blumenschine, R. J., Cavallo, J. A., & Capaldo, S. D. (1994). Competition for carcasses and early hominid behavioral ecology: A case study and conceptual framework. *Journal of Human Evolution*, 27(1–3), 197–213.

Bowlby, J. (1980). *Attachment and loss: Separation anxiety and anger* (vol. 2). n.p.: Basic Books.

Bråten, S. (1998). *Intersubjective communication and emotion in early ontogeny*. Cambridge: Cambridge University Press.

Brooks, R., & Meltzoff, A. N. (2002). The importance of eyes: How infants interpret adult looking behavior. *Developmental Psychology*, 38(6), 958–966.

———. (2008). Infant gaze following and pointing predict accelerated vocabulary growth through two years of age: A longitudinal, growth curve modeling study. *Journal of Child Language*, 35(1), 207–220.

Broom, R., 1938. The Pleistocene anthropoid apes of South Africa. *Nature*, 142, 377–379.

Brown, P.; et al. (2004). A new small-bodied hominin from the Late Pleistocene of Flores, Indonesia. *Nature*. 431(7012), 1055–1061.

Brown, R. (1970). *Psycholinguistics: Selected papers by Roger Brown*. New York: Free Press.

———. (1973). *A first language: The early stages* (illustrated 7th ed.). Cambridge, MA: Harvard University Press.

Bruner, J. S. (1983). *Child's talk: Learning to use language*. New York: Norton.

——. (1985). Child's talk: Learning to use language. *Child Language Teaching and Therapy*, 1(1), 11114.

Brunet, M., Beauvilain, A., Coppens, Y., Heintz, É., Moutaye, A. H. E, & Pilbeam, D. (1995). The first australopithecine 2,500 kilometres west of the Rift Valley (Chad). *Nature*. 378 (6554): 273–275.

Bunn, H. T., Kroll, E. M., Ambrose, S. H., Behrensmeyer, A. K., Binford, L. R., Blumenschine, R. J., Klein, R. G., McHenry, H. M., O'Brien, C. J.,& Wymer, J. J. (1986). Systematic butchery by Plio/Pleistocene hominids at Olduvai Gorge, Tanzania [and Comments and Reply]. *Current Anthropology*, 27(5), 431-452.

Burkart, J. M., Hrdy, S. B., & Van Schaik, C. P. (2009). Cooperative breeding and human cognitive evolution. *Evolutionary Anthropology*, 18, 175–186.

Carpenter, M., Akhtar, N., & Tomasello, M. (1998). Fourteen- through 18-month-old infants differentially imitate intentional and accidental actions. *Infant Behavior and Development*, 21(2), 315–330.

Carpenter, M., & Call, J. (2013). How joint is the joint attention of apes and human infant? In J. Metcalfe & H. Terrace (eds.), *Agency and joint attention*, 4961. Oxford: Oxford University Press.

Charman, T. (2003). Why is joint attention a pivotal skill in autism? *Philosophical Transactions of the Royal Society of London B*, 358(1430), 315–324.

Cheney, D. L., & Seyfarth, R. M. (1992). *How monkeys see the world: Inside the mind of another species*. Chicago: University of Chicago Press.

Chomsky, N. (1957). *Syntactical structures*. The Hague: Mouton.

——. (1959). A review of BF Skinner's *Verbal Behavior. Language*, 35(1), 26–58.

——. (1965). *Aspects of the theory of syntax*. Cambridge, MA: MIT Press.

——. (1972). *Language and mind*. New York: Harcourt Brace Jovanovich.

——. (1975). *Reflections on language*. New York: Pantheon.

——. (1980). Rules and representations. *Behavioral and Brain Sciences*, 3(1), 1–15.

——. (1988). *Language and problems of knowledge: The Managua lectures*. Cambridge, MA: MIT Press.

——. (2000). *New horizons in the study of language and mind*. Cambridge: Cambridge University Press.

——. Chomsky, N. (2005). *Some simple evo devo theses: How true might they be for language?* Presented at the Alice V. and David H. Morris Symposium on language and communication, Stonybrook University, New York.

——. (2010). Some simple evo devo theses: How true might they be for language? In R. K. Larson, V. Deprez, & H. Yamakido (eds.), *The evolution of human language*, 45–62. Cambridge: Cambridge University Press.

REFERENCES

Christiansen, M. H., & Kirby, S. (2003). Language evolution: Consensus and controversies. *Trends in Cognitive Sciences*, 7(7), 300–307.

Clark, H. (1996). *Using language*. Cambridge: Cambridge University Press.

Copeland, R., & Hall, H. V. (1976). Behavior modification in the classroom. *Progress in Behavior Modification*, 3(1), 45–78.

Coppens, Y. (1994). East side story: The origin of humankind. *Scientific American*, 270(5), 88–95.

Cullen, H. M., deMenocal, P. B., Hemming, S., Hemming, G., Brown, F., Guilderson, T., & Sirocko, F. (2000). Climate change and the collapse of the Akkadian empire: Evidence from the deep sea. *Geology*, 28(4), 379–382.

Dąbrowska, E. (2015). What exactly is Universal Grammar, and has anyone seen it? *Frontiers in Psychology*, 6, 852.

Dart R. A. (1925). *Australopithecus africanus*: the man-ape of South Africa. *Nature*, 115:195–199.

Darwin, C. (1859). *On the origin of species by means of natural selection*, ed. J. Carroll. (2003). Peterborough, UK: Broadview Press.

——. (1869). Letter to Alfred Russell Wallace, March 27, 1869. Darwin correspondence. https://www.darwinproject.ac.uk/entry-6684.

——. (1872). *The expression of the emotions in man and animals*. London: Clowes and Sons.

Davies, T. (2008). Environmental health impacts of East African Rift volcanism. *Environmental Geochemistry and Health*, 30, 325–338.

Deacon, T. W. (1997). *The symbolic species: The co-evolution of language and the brain*. New York: Norton.

Demenocal, P. B. (2004). African climate change and faunal evolution during the Pliocene–Pleistocene. *Earth and Planetary Science Letters*, 220(1), 3–24.

D'Errico, F., Henshilwood, C., Lawson, G., Vanhaeren, M., Tillier, A.-M., Soressi, M., . . . & Lakarra, J. (2003). Archaeological evidence for the emergence of language, symbolism, and music—an alternative multidisciplinary perspective. *Journal of World Prehistory*, 17(1), 1–70.

DeSilva, J., & Lesnik, J. (2006). Chimpanzee neonatal brain size: Implications for brain growth in *Homo erectus*. *Journal of Human Evolution*, 51(2), 207–212.

De Waal, B. M. F. (2011). What is an animal emotion? *Annals of the New York Acadamy of Sciences*, 1224(1), 191–206.

Dobzhansky, T. (1973). Nothing in biology makes sense except in the light of evolution. *American Biology Teacher*, 35(3), 125–129.

Dominguez-Rodrigo, M., Pickering, T. R., Semaw, S., & Rogers, M. J. (2005). Cutmarked bones from Pliocene archeological sites at Gona, Afar, Ethiopia. *Journal of Human Evolution, 48,* 109–121.

Dubois, E. (1896). On *Pithecanthropus erectus*: A transitional form between man and the apes. *Journal of the Anthropological Institute of Great Britain and Ireland, 25,* 240–255.

——. (1937). On the fossil human skulls recently discovered in Java and *Pithecanthropus erectus. Man, 37,* 1–7.

Dunbar, R. I. M. (1998). *Grooming, gossip, and the evolution of language.* Cambridge, MA: Harvard University Press.

Dunham, P., & Moore, C. (1995). *Joint attention: Its origins and role in development.* Hillsdale, NJ: Lawrence Erlbaum.

Eigsti, I.-M., de Marchena, A. B., Schuh, J. M., & Kelley, E. (2011). Language acquisition in autism spectrum disorders: A developmental review. *Research in Autism Spectrum Disorders, 5*(2), 681–691.

Evans, N., & Levinson, S. C. (2009). The myth of language universals: Language diversity and its importance for cognitive science. *Behavioral and Brain Sciences, 32*(5), 429–448.

Everett, D. L. (2012). *Language: The cultural tool.* New York: Pantheon.

——. (2017). *How language began. How Language Began the Story of Humanity's Greatest Invention.* New York: Liveright Publishing Corporation, a Division of W.W. Norton & Company.

Falk, D. (2004). Prelinguistic evolution in early hominins: Whence motherese? *Behavioral and Brain Sciences, 27,* 491–541.

Fleagle, J. (1998). *Primate adaptation and evolution* (2nd ed.). San Diego, CA: Academic Press. Reprinted with permission from Elsevier.

Folinsbee, R. E., Lipson, J., & Reynolds, J. H. (1956). Potassium-argon dating. *Geochimica et Cosmochimica Acta, 10*(1–2), 60–68.

Fortune, E. S., Rodríguez, C., Li, D., Ball, G. F., & Coleman, M. J. (2011). Neural mechanisms for the coordination of duet singing in wrens. *Science, 334*(6056), 666–670.

Fouts, R. S. (1975). Capacity for language in great apes. In R. Tuttle (ed.), *Socioecology and psychology of primates,* 371–390. Chicago: Mouton.

Fu, Q., Li, H., Moorjani, P., Jay, F., Slepchenko, S. M., Bondarev, A. . . . & Pääbo, S. (2014). Genome sequence of a 45,000-year-old modern human from western Siberia. *Nature, 514* (7523), 445–449.

Gardner, B. T. (1981). Project Nim: Who taught whom? *Contemporary Psychology, 26,* 425–426.

Gardner, H. (1985). *The mind's new science: A history of the cognitive revolution.* New York: Basic Books.

Gardner, R. A., & Gardner, B. T. (1969). Teaching sign language to a chimpanzee. *Science,* 165(894), 664–672.

Gardner, R. A., & Gardner, B. T., producers. (1973). *Teaching sign language to the chimpanzee: Washoe* [documentary]. No. 16802. University Park, PA: Psychological Cinema Register.

Gibbons, Ann (2007). *The first human: The race to discover our earliest ancestors.* New York: Anchor Books.

Gould, S. J. (1977). The return of hopeful monsters. *Natural History,* 86(6), 22–30.

Green, R. E., Malaspinas, A. S., Krause, J., Briggs, A. W., Johnson, P. L., Uhler, C. . . . & Pääbo, S. (2008). A complete Neandertal mitochondrial genome sequence determined by high-throughput sequencing. *Cell, 134*(3), 416–426.

Greenfield, P. M. (1991). Language, tools and brain: The ontogeny and phylogeny of hierarchically organized sequential behavior. *Behavioral and Brain Sciences,* 14(4), 577–595.

Guatelli-Steinberg, D. (2003). Macroscopic and microscopic analyses of linear enamel hypoplasia in Plio-Pleistocene South African hominins with respect to aspects of enamel development and morphology. *American Journal of Physical Anthropology,* 120, 309–322.

Hadhazy, A. (2012). Do pheromones play a role in our sex lives? *Scientific American* (blog post). https://www.scientificamerican.com/article/pheromones -sex-lives/.

Haile-Selassie, Y. (2001). Late Miocene hominids from the Middle Awash, Ethiopia. *Nature,* 412(6843): 178–181.

Hare, B., Call, J., Agnetta, B., & Tomasello, M. (2000). Chimpanzees know what conspecifics do and do not see. *Animal Behaviour,* 59, 771–785.

Hare, B., Call, J., & Tomasello, M. (2001). Do chimpanzees know what conspecifics know? *Animal Behaviour,* 61, 139–151.

Harlow, H. F. (1949). The formation of learning sets. *Psychological Review,* 56(1), 51.

Harrar, L. (Writer). (1984). Signs of the apes, songs of the whales. WGBH (Television station: Boston, MA): Time Life Video.

Hauser, M. D., Chomsky, N., & Fitch, W. T. (2002). The faculty of language: What is it, who has it, and how did it evolve? *Science,* 298(5598), 1569–1579.

Hauser, M., Yang, C., Berwick, R., Tattersall, I., Ryan, M., Watumull, J., Chomsky, N., & Lewontin, R. (2014). The mystery of language evolution. *Frontiers in Psychology*, 5, 1–12.

Hayes, C. (1951). *The ape in our house*. New York: Harper.

Hayes, C., & Hayes, K. (1951). The intellectual development of a home-raised chimpanzee. *Proceedings of the American Philosophical Society*, 95, 105–109.

Heine, B., & Kuteva, T. (2007). *The genesis of grammar: A reconstruction* (vol. 9). New York: Oxford University Press.

Herrnstein, R. J. (1985). Riddles of natural categorization. *Philosophical Transactions of the Royal Society of London*, 308 B, 129–144.

Higgins, E. T. (2016). Shared-reality development in childhood. *Perspectives on Psychological Science*, 11(4), 466–495.

Hinde, R. A. (1987). *Individuals, relationships & culture: Links between ethology and the social sciences*. New York: Cambridge University Press.

Hinde, R. A., & Stevenson-Hinde, J. (1987). Interpersonal relationships and child development. *Developmental Review*, 7(1), 1–21.

Hobaiter, C., & Byrne, R. W. (2014). The meanings of chimpanzee gestures. *Current Biology*, 24(14), 1596–1600.

Hobson, P. (2002). *The cradle of thought: Exploring the origins of thinking*. London: Macmillan.

Hockett, C. F., & Altmann, S. (1968). A note on design features. In T. A. Sebeok (ed.), *Animal communication: Techniques of study and results of research*, 6172. Bloomington: Indiana University Press.

Hoffmann, D. L., Standish, C. D., García-Diez, M., et al. (2018). U-Th dating of carbonate crusts reveals Neandertal origin of Iberian cave art. *Science*, 359(6378), 912–915.

Hrdy, S. B. (2009). *Mothers and others: The evolutionary origins of mutual understanding*. Cambridge, MA: Belknap Press.

Hurford, J. (2007). *The origins of meaning*. Oxford: Oxford University Press.

Huxley, T. H. (1887). On the reception of the 'Origin of Species.' In F. Darwin (ed.), *The life and letters of Charles Darwin* (1904). New York: Appleton.

Jaanusson, V. (2007). Balance of the head in hominid evolution. *Lethaia*, 20, 165–176.

Jackendoff, R. (1999). Possible stages in the evolution of the language capacity. *Trends in Cognitive Sciences*, 3(7), 272–279.

——. (2002). *Foundations of language: Brain, meaning, grammar, evolution*. New York: Oxford University Press.

Jacob, F. (1982). *The Possible and the Actual*. New York: Pantheon.

Jaffe, J., & Anderson, S. (1979). Communication rhythms and the evolution of language. In A. Siegman & S. Feldstein (eds.), *Of speech and time: Temporal speech patterns and interpersonal conflicts*, 17–22. Hillsdale, NJ: Lawrence Erlbaum.

Jaffe, J., Beebe, B., Feldstein, S., Crown, C. L., & Jasnow, M. D. (2001). Rhythms of dialogue in infancy: Coordinated timing in development. *Monographs of the Society for Research in Child Development* 66(2), iviii, 1132.

Jensen, G., Altschul, D., & Terrace, H. (2013). Monkeys would rather see and do: Preference for agentic control in rhesus macaques. *Experimental Brain Research*, 229(3), 429–442.

Johanson, D.C. (2009). Lucy (*Australopithecus afarensis*). In Michael Ruse; Joseph Travis, *Evolution: The First Four Billion Years*, 693–697. Cambridge, MA: Belknap Press.

Jones, J. (2018, February 23). So Neanderthals made abstract art? This astounding discovery humbles every human. *The Guardian*. https://www.theguardian.com/artanddesign/2018/feb/23/neanderthals-cave-art-spain-astounding-discovery-humbles-every-human.

Kaminski, J., Call, J., & Fischer, J. (2004). Word learning in a domestic dog: Evidence for "fast mapping." *Science*, 304, 1682–1683.

Kaminski, J., Riedel, J., Call, J., & Tomasello, M. (2005). Domestic goats, *Capra hircus*, follow gaze direction and use social cues in an object choice task. *Animal Behaviour*, 69(1), 11–18.

Kaye, D. (2014, July 1). Human see, human do: A complete history of "Planet of the Apes." *Rolling Stone*. https://www.rollingstone.com/movies/news/human-see-human-do-a-complete-history-of-planet-of-the-apes-20140701.

Kellogg, W. N., & Kellogg, L. A. (1933). *The ape and the child: A study of environmental influence and its behavior*. New York: McGraw-Hill.

Kenneally, C. (2007). *The first word: The search for the origins of language*. New York: Viking.

King, W. (1864.) The reputed fossil man of the Neanderthal. *Quarterly Review of Science*, 1:88–97.

Köhler, W. (1925). *The mentality of apes*. New York: Harcourt, Brace & World.

Kohts, N. (1935). *Infant ape and human child*. Moscow, Russia: Museum Darwinianum.

Laland, K. N. (2017). *Darwin's unfinished symphony: How culture made the human mind*. Princeton, NJ: Princeton University Press.

Langmore, N. E. (1998). Functions of duet and solo songs of female birds. *Trends in Ecology & Evolution*, 13(4), 136–140.

Larson, S. G. (2009). Evolution of the hominin shoulder: Early *Homo*. In F. E. Grine, J. G. Fleagle, & R. E. Leakey (eds.), *The first humans: Origin and early evolution of the genus Homo*, 6575. New York: Springer Science + Business.

Leakey, L. S. B. (1959). A new fossil from Olduvai. *Nature*, 184, 491–494.

Leakey, M. G., Feibel, C. S., McDougall, I., Walker, A. (1995). New four-million-year-old hominid species from Kanapoi and Allia Bay, Kenya. *Nature*, 376, 565–571.

Leakey M. G., Spoor F., Brown F., Gathogo P. N., Kiarie C., Leakey L. N., et al. (2001). New hominin genus from eastern Africa shows diverse middle Pliocene lineages. *Nature*, 410, 433–440.

Leakey, L. S. B., Tobias, P. V., & Napier, J. R. (1964). A new species of the genus *Homo* from Olduvai Gorge. *Nature*, 202, 7–9.

Lehmicke, A. (artist). n.d. Cut marks vs. tool marks. Retrieved from http://anthropologylabs.umn.edu/digital/CaseDisplay/pages/predation/index.html.

Lewis, B., Jurmain, R., & Kilgore, L. (2006). *Understanding physical anthropology and archaeology* (9th ed.). Belmont, CA: Wadsworth. Republished with permission of South-Western-College Publishing, a division of Cengage Learning

Lieberman, P. (1975). *On the origins of language: An introduction to the evolution of human speech*. New York: Macmillan.

Luiselli, J. K., Putnam, R. F., Handler, M. W., & Feinberg, A. B. (2005). Whole-school positive behaviour support: Effects on student discipline problems and academic performance. *Educational Psychology*, 25(2–3), 183–198.

MacNamara, J. (1972). Cognitive basis of language learning in infants. *Psychological Review*, 79(1), 1–13.

Manzi, G; Mallegni, F; Ascenzi, A (August 2001). A cranium for the earliest Europeans: Phylogenetic position of the hominid from Ceprano, Italy. *Proc. Natl. Acad. Sci. U.S.A.* 98 (17), 10011–10016.

Markson, L., & Bloom, P. (1997). Evidence against a dedicated system for word learning in children. *Nature*, 385, 813–815.

Maslin, M., Pancost, R., Wilson, K., Lewis, J., & Trauth, M. (2012). Three and half million year history of moisture availability of South West Africa: Evidence from ODP site 1085 biomarker records. *Palaeogeography, Palaeoclimatology, Palaeoecology*, 317, 41–47.

REFERENCES

Matticus78 (artist). (2006). *Diagram of eye evolution* [digital image]. Wikimedia Commons. Retrieved from https://commons.wikimedia.org/wiki/File: Diagram_of_eye_evolution.svg.

Mayr, E. (2001). *What evolution is.* New York: Basic Books.

McBrearty, S., & Jablonski, N. G. (2005). First fossil chimpanzee. *Nature*, 437(7055), 105–108.

Meltzoff, A. N., & Moore, K. M. (1977). Imitation of facial and manual gestures by human neonates. *Science*, 198(4312), 75–78.

———. (1989). Imitation in newborn infants: Exploring the range of gestures imitated and the underlying mechanisms. *Developmental Psychology*, 25(6), 954–962.

———. (1994). Imitation, memory, and the representation of persons. *Infant Behavior and Development*, 17, 83–99.

———. (1998). Infant intersubjectivity: Broadening the dialogue to include imitation, identity and intention. In S. Bråten (ed.), *Intersubjective communication and emotion in early ontogeny* (2006). Cambridge: Cambridge University Press.

Metcalfe, J., & Terrace, H. S. (2013). *Agency and joint attention.* New York: Oxford University Press.

Miles, H. L. (1983). Apes and language: The search for communicative competence. In J. de Luce & H. T. Wilder (eds.), *Language in primates: Perspectives and implications* (pp. 43–61). New York: Springer-Verlag.

Mithen, S., Morley, I., et al. (2006). The singing Neanderthals: The origins of music, language, mind and body. *Cambridge Archaeological Journal*, 16(1), 97–112.

Miyagawa, S., Lesure, C., & Nóbrega, V. A. (2018). Cross-modality information transfer: A hypothesis about the relationship among prehistoric cave paintings, symbolic thinking, and the emergence of language. *Frontiers in Psychology*, 9 (February 20): 115.

Mounier, A., Marchal, F., Condemi, S. (2009). Is *Homo heidelbergensis* a distinct species? New insight on the Mauer mandible. *Journal of Human Evolution*, 56, 219–246.

Morales, M., Mundy, P., Delgado, C., Yale, M., Messinger, D., Neal, R., & Schwartz, H. (2000). Responding to joint attention across the 6- through 24-month age period and early language acquisition. *Journal of Applied Developmental Psychology*, 21(3), 283–298.

Müller, M. (1862). *The science of lanuage.* New York: Charles Scribner.

Mundy, P., Block, J., Delgado, C., et al. (2007). Individual differences and the development of joint attention in infancy. *Child Development*, 78, 938–954.

Mundy, P., & Newell, L. (2007). Attention, joint attention and social cognition. *Current Directions in Psychological Science*, 16(5), 269–274.

Mundy, P., Sigman, M., & Kasari, C. (1990). A longitudinal study of joint attention and language development in autistic children. *Journal of Autism and Developmental Disorders*, 20, 115–128.

Murray, L., & Trevarthen, C. (1985). Emotional regulation of interactions between two-month-olds and their mothers. In T. M. Field & N. A. Fox (eds.), *Social perception in infants*. Norwood, NJ: Ablex.

Nagy, E., & Molnar, P. (1994). Homo imitans or homo provocans? *International Journal of Psychophysiology*, 18(2), 128.

Napier, J. (1967). The Antiquity of Human Walking. *Scientific American*, 216(4), 56–67.

Nelson, C. A., Zeanah, C. H., Fox, N. A., et al. (2007). Cognitive recovery in socially deprived young children: The Bucharest Early Intervention Project. *Science*, 318(5858), 1937–1940.

Pante, M. C., Scott, R. S., Blumenschine, R. J., & Capaldo, S. D. (2015). Revalidation of bone surface modification models for inferring fossil hominin and carnivore feeding interactions. *Quaternary International*, 355, 164–168.

Patterson, F. G. (1981). Ape language. *Science*, 211, 86–87.

Pavlov, I. P. (1927). *Conditioned reflexes*. Trans. G. V. Anrep. New York: Oxford University Press.

Pfungst, O. (1911). *Clever Hans*. Trans. C. Stumpf & C. L. Rahn. Oxford: Holt.

Pilley, J. W. & Reid, A. K. (2011). Border collie comprehends object names as verbal referents. *Behavioural Processes*, 86(2), 184–195.

Pinker, S. (1994). *The language instinct*. New York: William Morrow.

Pobiner, B. (2016). Meat-eating among the earliest humans. *American Scientist*, 104(2), 110.

Potts, R., (1996). Evolution and climate variability. *Science* 273: 922–923.

——. (2012). Hominin evolution in settings of strong environmental variability. *Quaternary Science Reviews*, 73, 1–13.

Povinelli, D. J., & Vonk, J. (2003). Chimpanzee minds: Suspiciously human? *Trends in Cognitive Science*, 7(4), 157–160.

Powell, A. (2010, October 5). 'Breathtakingly awful.' Retrieved from https://news.harvard.edu/gazette/story/2010/10/breathtakingly-awful/.

Premack, A. J., & Premack, D. (1972). Teaching language to an ape. *Scientific American*, 227 (4): 1072–1092.

Premack, D. (1976). *Intelligence in ape and man*. Hillsdale, NJ: Lawrence Erlbaum.

——. (1986). *Gavagai*. Cambridge, MA: MIT Press.

Progovac, L. (2015). *Evolutionary syntax.* Oxford: Oxford University Press.

Quine, W. V. O. (1960). *Word and object.* Cambridge, MA: MIT Press.

Range, F., & Virányi, Z. (2011). Development of gaze following abilities in wolves (*Canis lupis*). *PLoS ONE*, 6(2), 1–9.

Reader, J. (2011). *Missing links: In search of human origins.* Oxford: Oxford University Press.

Rightmire, G. P. (2013). *Homo erectus* and middle Pleistocene hominins: Brain size, skull form, and species recognition. *Journal of Human Evolution*, 65(3), 109–123.

Ristau, C. A., & Robbins, D. (1982). Language in the great apes: A critical review. *Advances in the Study of Behavior*, 12, 141–255.

Roitblat, H. L., Bever, T. G., & Terrace, H. S. (1984). *Animal cognition.* Hillsdale, NJ: Lawrence Erlbaum.

Rosenberg, K., & Trevathan, W. (2002). Birth, obstetrics and human evolution. *An International Journal of Obstetrics and Gynaecology*, 109(11), 1199–1206.

Rumbaugh, D. M. (1977). *Language learning by a chimpanzee: The Lana project.* New York: Academic Press.

Rumbaugh, D. M. (2013). *With apes in mind: Emergents, communication and competence . . . So that together we might learn of language.* D. M. Rumbaugh.

Rumbaugh, D. M., Gill, T. V., & Von Glasersfeld, E. C. (1973). Reading and sentence completion by a chimpanzee (Pan). *Science*, 182(4113), 731–733.

Rutter, M. (1998). Developmental catch-up, and deficit, following adoption after severe global early privation. *Journal of Child Psychology and Psychiatry*, 39(4), 465–476.

Sahle, Y., El Zaatari, S., & White, T. (2017). Hominid butchers and biting crocodiles in the African Plio-Pleistocene. *PNAS*, 114(50), 13164–13169.

Sánchez-Quinto, F., Botigué, L. R., Civit, S., Arenas, C., Avila-Arcos, M. C., Bustamante, C. D., Comas, D, & Lalueza-Fox, C. (2012). North African populations carry the signature of admixture with neandertals. *PLoS ONE*, 7(10).

Saussure, F. (1916/1959). *Course in general linguistics* (ed. C. Bally, A. Sechehaye, & A. Riedlinger, trans. R. Harris). London: Duckworth.

Savage-Rumbaugh, E. S. (1994). *Kanzi: The ape at the brink of the human mind.* New York: Wiley.

Savage-Rumbaugh, E. S., Murphy, J., Sevcik, R., Brakke, K. E., Williams, S. L., & Rumbaugh, D. M. (1993). Language comprehension in ape and child. *Monographs of the Society for Research in Child Development*, 58(3–4), 1–256.

Savage-Rumbaugh, E. S., Rumbaugh, D. M., & Boysen, S. (1978). Linguistically mediated tool use and exchange by chimpanzees (*Pan troglodytes*). *Behavioral and Brain Sciences*, 1(4), 539–554.

Savage-Rumbaugh, E. S., Rumbaugh, D. M., Smith, S. T., & Lawson, J. (1980). Reference: The linguistic essential. *Science*, 210, 922–925.

Scaife, M., & Bruner, J. (1975). The capacity for joint visual attention in the infant. *Nature*, 253, 265–266.

Schilbach, L., Timmermans, B., Reddy, V., Costall, A., Bente, G., Schlicht, T., & Vogeley, K. (2013). Toward a second-person neuroscience. *Behavioral and Brain Sciences, 36(4)*, 393–414.

Schloegl, C., Kotrschal, K., & Bugnyar, T. (2007). Gaze following in common ravens, *Corvus corax*: Ontogeny and habituation. *Animal Behaviour*, 74(4), 769–778.

Schroeder, B. (writer). (1978). *Koko: A talking gorilla* [documentary]. Home Vision.

Schusterman, R., & Kastak, D. (1998). Functional equivalence in a California sea lion: Relevance to animal social and communicative interactions. *Animal Behavior*, 55, 1087–1095.

Senut, B. (2006). The 'East Side Story' twenty years later. *Transactions of the Royal Society of South Africa*, 61(2), 103–109.

Senut, B., Pickford, M., Gommery, D., Mein, P., Cheboi, K., & Coppens, Y. (2001). First hominid from the Miocene (Lukeino Formation, Kenya). *Comptes Rendus de l'Académie des Sciences*, 332(2): 137–144.

Shefelbine, S., Tardieu, C., & Carter, D. R. (2002). Development of the femoral bicondylar angle in hominid bipedalism. *Bone*, 30(5), 765–770. Reprinted with permission from Elsevier.

Shively, C. (1985). The evolution of dominance hierarchies in nonhuman primate society. In S. L. Ellyson & J. F. Dovidio (eds.), *Power, dominance, and nonverbal behavior*, 67–87. New York: Springer.

Shreeve, J. (1996). Sunset on the savanna. *Discover*, 17(7), 116–125.

Silk, J. B., Brosnan, S. F., Henrich, J., et al. (2013). Chimpanzees share food for many reasons: The role of kinship, reciprocity, social bonds and harassment on food transfers. *Animal Behaviour*, 85(5), 941–947.

Simpson, G. G. (1945). The principles of classification and a classification of mammals. *Bulletin of the American Museum of Natural History*, 85.

Skinner, B. F. (1938). *The behavior of organisms*. New York: Appleton-Century-Crofts.

——. (1957). *Verbal behavior*. New York: Appleton-Century-Crofts.

———. (1959). *Cumulative record*. New York: Appleton-Century-Crofts.

———. (1960). Pigeons in a pelican. *American Psychologist, 15*(1), 28–37.

Slon, V., Mafessoni, F., Vernot, B., de Filippo, C., Grote, S., Viola, B., H. . . . & Pääbo, S. (2018). The genome of the offspring of a Neanderthal mother and a Denisovan father. *Nature*, 561, 113–116.

Spencer, H. (1886). *The principles of psychology*. New York: Appleton.

Stern, D. (1971). A microanalysis of the mother-infant interaction. *Journal of the American Academy of Child Psychiatry*, 10(3), 501–517.

———. (1985). *The interpersonal world of the infant*. New York: Basic Books.

Stoddard, P. K., & Markham, M. R. (2008). Signal cloaking by electric fish. *BioScience*, 58(5), 415–425.

Stokoe, W. C., Casterline, D. C., & Croneberg, C. G. (1965). *A dictionary of American Sign Language on linguistic principles*. Washington, DC: Gallaudet Press.

Studdert-Kennedy, M., & Terrace, H. (2017). In the beginning. *Journal of Language Evolution*, 10(1093), 114–125.

Sundberg, M. L., & Michael, J. (2001). The benefits of Skinner's analysis of verbal behavior for children with autism. *Behavior Modification*, 25(5), 698–724.

Swisher III, C. C., Curtis, G. H., & Lewin, R. (2001). *Java man: How two geologists changed our understanding of human evolution*. Chicago: University of Chicago Press.

Tanner, N., Jensen, G., Ferrera, V. P., & Terrace, H. S. (2017). Inferential learning of serial order of perceptual categories by rhesus monkeys (*Macaca mulatta*). *Journal of Neuroscience*, 37(26), 6268–6276.

Teaford, M., & Ungar, P. S. (2000). Diet and the evolution of the earliest human ancestors. *PNAS*, 97(25), 13506–13511.

Teglas, E., Gergely, A., Kupan, K., Miklosi, A., & Topal, J. (2012). Dogs' gaze following is tuned to human communicative signals. *Current Biology*, 22, 209–212.

Tennyson, A. L. (1850). *In Memoriam*. London: Edward Moxon.

Terrace, H. (1963a). Discrimination learning with and without errors. *Journal of Experimental Analysis of Behavior*, 6, 1–27.

———. (1963b). Errorless transfer of a discrimination across two continua. *Journal of Experimental Analysis of Behavior*, 6, 223–232.

———. (1966). Stimulus control. In W. K. Honig (ed.), *Operant behavior areas of research and application*, 291–344. Englewood Cliffs, NJ: Prentice Hall.

———. (1979). *Nim*. New York: Knopf.

———. (1981). Reply to Bindra and Patterson. *Science*, 211, 87–88.

——. (1985). In the beginning was the name. *American Psychologist*, 40, 1011–1028.

——. (2005). The simultaneous chain: A new approach to serial learning. *Trends in Cognitive Sciences*, 9(4), 202–210.

——. (2012). Schoff Lectures: 1. Mind the Gap: *Why Two Minds Are Better Than One: The Evolution of Words.*, 2. Intelligence of Non-Human Primates., 3. Development of Non-Verbal & Uniquely Human Behavior During an Infant's First Year. November, 2012. Columbia University, New York.

——. (2019). OriginOfWords@Book.Psych.Columbia.edu. Retrieved from OriginOfWords@Book.Psych.Columbia.edu.

Terrace, H. S., Petitto, L. A., Sanders, R. J., & Bever, T. G. (1979). Can an ape create a sentence? *Science*, 206(4421), 891–902.

Terrace, H. S., Son, L., & Brannon, E. (2003). Serial expertise of rhesus macaques. *Psychological Sciences*, 14, 66–73.

Terrace, H. S., & Studdert-Kennedy, M. (2015, November 3). Commentary on "The Mystery of Language Evolution." *Language Log*. Guest post filed by M. Lieberman. http://languagelog.ldc.upenn.edu/nll/?p=22011.

Thompson, C. R., & Church, R. M. (1980). An explanation of the language of a chimpanzee. *Science*, 208, 313–314.

Thompson, R. K., & Herman, L. M. (1977). Memory for lists of sounds by the bottle-nosed dolphin: Convergence of memory processes with humans? *Science*, 195(4277), 501–503.

Tomasello, M. (1999). *The cultural origins of human cognition.* London: Harvard University Press.

Tomasello, M., Carpenter, M., Call, J., Behne, T., & Moll, H. (2005). Understanding and sharing intentions: The origins of cultural cognition. *Behavioral and Brain Sciences*, 28, 675–735.

Tomonaga, M., Tanaka, M., Matsuzawa, T., et al. (2004). Development of social cognition in infant chimpanzees (*Pan troglodytes*): Face recognition, smiling, gaze, and the lack of triadic interactions. *Japanese Psychological Research*, 46(3), 227–235.

Tottenham, N., Hare, T. A., Quinn, B. T., et al. (2010). Prolonged institutional rearing is associated with atypically large amygdala volume and difficulties in emotion regulation. *Developmental Science*, 13(1), 46–61.

Trevarthen, C. (1977). Descriptive analysis of infant communicative behavior. In H. R. Schaffer (ed.), *Studies in mother-infant interaction*, 227270. London: Academic Press.

——. (1980). The foundations of intersubjectivity: Development of interpersonal and cooperative understanding in infants. In D. R. Olson (ed.), *The social foundations of language and thought*, 316–342. New York: Norton.

——. (1998). The concept and foundations of infant intersubjectivity. In S. Bråten (ed.), *Studies in emotion and social interaction, 2nd series. Intersubjective communication and emotion in early ontogeny*, 15–46. New York: Cambridge University Press.

Trevarthen, C., & Aitken, K. J. (2001). Infant intersubjectivity: Research, theory, and clinical applications. *Journal of Child Psychology and Psychiatry*, 42(1), 3–48.

Tronick, E. Z., Als, H., & Adamson, L. (1979). Structure of early face-to-face communicative interactions. In M. Bullowa (ed.), *Before speech: The beginning of interpersonal communication*, 349372. Cambridge: Cambridge University Press.

van Ginneken, V., Van Meerveld, A., Wijgerde, T., Verheij, E., de Vries, E. & van der Greef, J. (2017). Hypothesis Hunter-prey correlation between migration routes of African buffaloes and early hominids: Evidence for the " Out of Africa " hypothesis. *Integrative Molecular Medicine*, 4, 1–5.

von Frisch, K. (1967). Honeybees: Do they use direction and distance information provided by their dancers? *Science*, 158(3804), 1072–1077.

Wade, N. (1980). Does man alone have language? Apes reply in riddles, and a horse says neigh. *Science*, 208, 1349–1351.

Wallace, A. R. (1870). *Contributions to the theory of natural selection: A series of essays* (2nd ed.). London: Macmillan.

Walker, A. C., Leakey, R. E., Harris, J. M., & Brown, F. H. (1986). 2.5-Myr Australopithecus boisei from west of Lake Turkana, Kenya. *Nature*, 322, 517–522.

Walter, R. C. (1997). Potassium-argon/argon-argon dating methods. In *Chronometric dating in archaeology*, 97126. Boston: Springer.

Warneken, F., Chen, F., & Tomasello, M. (2006). Cooperative activities in young children and chimpanzees. *Child Development*, 77(3), 640–663.

Warneken, F., & Tomasello, M. (2006). Altruistic helping in human infants and young chimpanzees. *Science*, 311(5765), 1301–1303.

Washburn, S. L., & Lancaster, C. S. (1968). The evolution of hunting. In R. B. Lee & I. DeVore (eds.), *Man the hunter*, 293–303. Chicago: Aldine.

Wasserman, E. A., & Zentall, T. R. (2012). *The Oxford handbook of comparative cognition*. New York: Oxford University Press.

White, T. D., Suwa, G., & Asfaw, B. (1994). Australopithecus ramidus, a new species of early hominid from Aramis, Ethiopia. *Nature*, 371 (6495): 306–312.

Wilkes-Gibbs, D., & Clark, H. H. (1992). Coordinating beliefs in conversation. *Journal of Memory & Language*, 31, 183–194.

Wilkinson, A., Mandl, I., Bugnyar, T., & Huber, L. (2010). Gaze following in the red-footed tortoise. *Animal Cognition*, 13(5), 765–769.

Wilson, E. O. (1962). Chemical communication in the fire ant *Solenopsis saevissima*. *Animal Behavior*, 10(1–2), 134–138.

Windsor, J., Glaze, L. E., Koga, S. F., & Bucharest Early Intervention Project Core Group. (2007). Language acquisition with limited input: Romanian institution and foster care. *Journal of Speech, Language, & Hearing Research*, 50(5), 1365–1381.

Wong, R. Y., & Hopkins, C. D. (2007). Electrical and behavioral courtship displays in the mormyrid fish *Brienomyrus brachyistius*. *Journal of Experimental Biology*, 210(13), 2244–2252.

Wood, B., & Collard, M. (2001). The meaning of Homo. *Ludus Vitalis*, IX, 63–74.

Wood, B., & Harrison, T. (2011). The evolutionary context of the first hominins. *Nature*, 470(7334), 347–352.

Wray, A. (1998). Protolanguage as a holistic system for social interaction. *Language & Communication*, 18(1), 47–67.

Wyatt, T. D. (2015). How animals communicate via pheromones. *American Scientist*, 103(2), 114–121.

Wynne, C. (2007). Aping language. *Skeptics Society*, 13(5), 10–14.

Yang, C. (2013). Ontogeny and phylogeny of language. *Proceedings of the National Academy of Sciences*, 110(16), 6324–6327.

Yerkes, R. M. (1925). *Almost human*. London: Century/Random House UK.

INDEX